W9-BYQ-145

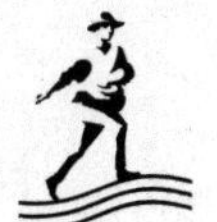

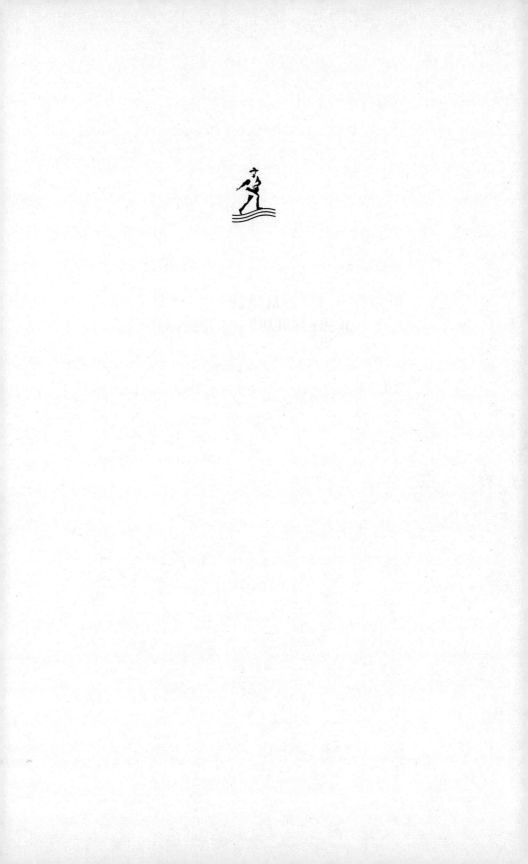

ALSO BY
JIMMY SONI AND ROB GOODMAN

Rome's Last Citizen:
The Life and Legacy of Cato, Mortal Enemy of Caesar

A MIND AT PLAY

How Claude Shannon Invented the Information Age

JIMMY SONI
AND
ROB GOODMAN

SIMON & SCHUSTER
New York London Toronto Sydney New Delhi

Simon & Schuster
1230 Avenue of the Americas
New York, NY 10020

First Simon & Schuster hardcover edition July 2017

Interior design by Lewelin Polanco

Manufactured in the United States of America

3 5 7 9 10 8 6 4 2

Library of Congress Cataloging-in-Publication Data

Names: Soni, Jimmy. | Goodman, Rob.
Title: A mind at play : how Claude Shannon invented the information age / by Jimmy Soni and Rob Goodman.
Description: New York : Simon & Schuster, 2017. | Includes bibliographical references and index.
Identifiers: LCCN 2016050944 (print) | LCCN 2016051979 (ebook) | ISBN 9781476766683 (hardcover : alk. paper) | ISBN 9781476766690 (trade pbk. : alk. paper) | ISBN 9781476766706 (ebook)
Subjects: LCSH: Shannon, Claude Elwood, 1916–2001. | Mathematicians—United States—Biography. | Electrical engineers—United States—Biography. | Information theory.
Classification: LCC QA29.S423 S66 2017 (print) | LCC QA29.S423 (ebook) | DDC 003/.54092 [B] —dc23
LC record available at https://lccn.loc.gov/2016050944

ISBN 978-1-4767-6668-3
ISBN 978-1-4767-6670-6 (ebook)

For Abigail, born 6/22/15

—R.G.

For Meagan

—J.S.

Contents

Geniuses are the luckiest of mortals because what they must do is the same as what they most want to do and, even if their genius is unrecognized in their lifetime, the essential earthly reward is always theirs, the certainty that their work is good and will stand the test of time. One suspects that the geniuses will be least in the Kingdom of Heaven—if, indeed, they ever make it; they have had their reward.

—**W. H. AUDEN**

INTRODUCTION

The thin, white-haired man had spent hours wandering in and out of meetings at the International Information Theory Symposium in Brighton, England, before the rumors of his identity began to proliferate. At first the autograph seekers came in a trickle, and then they clogged hallways in long lines. At the evening banquet, the symposium's chairman took the microphone to announce that "one of the greatest scientific minds of our time" was in attendance and would share a few words—but once he arrived onstage, the thin, white-haired man could not make himself heard over the peals of applause.

And then finally, when the noise had died down: "This is—ridiculous!" Lacking more to say, he removed three balls from his pocket and began to juggle.

After it was over, someone asked the chairman to put into perspective what had just happened. "It was," he said, "as if Newton had showed up at a physics conference."

It was 1985, and the juggler's work was long over, and just beginning. It had been nearly four decades since Claude Elwood Shannon published "the Magna Carta of the Information Age"—invented, in a single stroke, the idea of information. And yet the world his idea had made possible was only just coming into being. Now we live immersed in that world, and every email we have ever sent, every DVD and sound file we have ever played, and every Web page we have ever loaded bears a debt to Claude Shannon.

It was a debt he was never especially keen to collect. He was a man immune to scientific fashion and insulated from opinion of all kinds, on all subjects, even himself, especially himself; a man of closed doors and long silences, who thought his best thoughts in spartan bachelor apartments and empty office buildings. A colleague called Shannon's information theory "a bomb." It was stunning in its scope—he had conceived of a new science nearly from scratch—and stunning in its surprise—he had gone years barely speaking a word of it to anyone.

Of course, information existed before Shannon, just as objects had inertia before Newton. But before Shannon, there was precious little sense of information as an idea, a measurable quantity, an object fitted out for hard science. Before Shannon, information was a telegram, a photograph, a paragraph, a song. After Shannon, information was entirely abstracted into bits. The sender no longer mattered, the intent no longer mattered, the medium no longer mattered, not even the meaning mattered: a phone conversation, a snatch of Morse telegraphy, a page from a detective novel were all brought under a common code. Just as geometers subjected a circle in the sand and the disc of the sun to the same laws, and as physicists subjected the sway of a pendulum and the orbits of the planets to the same laws, Claude Shannon made our world possible by getting at the essence of information.

It is a puzzle of his life that someone so skilled at abstracting his way past the tangible world was also so gifted at manipulating it. Shannon was a born tinkerer: a telegraph line rigged from a barbed-wire fence, a makeshift barn elevator, and a private backyard trolley tell the story of his small-town Michigan childhood. And it was as an especially advanced sort of tinkerer that he caught the eye of Vannevar Bush—soon to become the most powerful scientist in America and Shannon's most influential mentor—who brought him to MIT and charged him with the upkeep of the differential analyzer, an analog computer the size of a room, "a fearsome thing of shafts, gears, strings, and wheels rolling on disks" that happened to be the most advanced thinking machine of its day.

Shannon's study of the electrical switches directing the guts of that mechanical behemoth led him to an insight at the foundation of our digital age: that switches could do far more than control the flow of electricity through circuits—that they could be used to evaluate any logical statement we could think of, could even appear to "decide." A series of binary choices—on/off, true/false, 1/0—could, in principle, perform a passable imitation of a brain. That leap, as Walter Isaacson put it, "became the basic concept underlying all digital computers." It was Shannon's first great feat of abstraction. He was only twenty-one.

A career that launched with "possibly the most important, and also the most famous, master's thesis of the century" brought him into contact and collaboration with thinkers like Bush, Alan Turing, and John von Neumann: all, like Shannon, founders of our era. It brought him into often-reluctant cooperation with the American defense establishment and into arcane work on cryptography, computer-controlled gunnery, and the encrypted transatlantic phone line that connected Roosevelt and Churchill in the midst of world war. And it brought him to Bell Labs, an industrial R&D operation that considered itself less an arm of the phone company than a home for "the operation of genius." "People did very well at Bell Labs," said one of Shannon's colleagues, "when they did what others thought was impossible." Shannon's choice of the impossible was, he wrote, "an analysis of some of the fundamental properties of general systems for the transmission of intelligence, including telephony, radio, television, telegraphy, etc."—systems that, from a mathematical perspective, appeared to have nothing essential in common until Shannon proved that they had everything essential in common. It would be his second, and greatest, feat of abstraction.

Before the publication of his "Mathematical Theory of Communication," scientists could track the movement of electrons in a wire, but the possibility that the very idea they stood for could be measured and manipulated *just as objectively* would have to wait until it was proved by Shannon. It was summed up in his recognition that all information, no matter the source, the sender, the recipient, or the meaning, could be efficiently represented by a sequence of *bits:* information's fundamental unit.

Before the "Mathematical Theory of Communication," a century of common sense and engineering trial and error said that noise—the physical world's tax on our messages—had to be lived with. And yet Shannon proved that noise could be defeated, that information sent from Point A could be received with perfection at Point B, not just often, but essentially always. He gave engineers the conceptual tools to digitize information and send it flawlessly (or, to be precise, with an arbitrarily small amount of error), a result considered hopelessly utopian up until the moment Shannon proved it was not. Another engineer marveled, "How he got that insight, how he even came to believe such a thing, I don't know."

That insight is embedded in the circuits of our phones, our computers, our satellite TVs, our space probes still tethered to the earth with thin cords of 0's and 1's. In 1990, the Voyager 1 probe turned its camera back on Earth from the edge of the solar system, snapped a picture of our planetary home reduced in size to less than a single pixel—to what Carl Sagan called "a mote of dust suspended in a sunbeam"—and transmitted that picture across four billion miles of void. Claude Shannon did not write the code that protected that image from error and distortion, but, some four decades earlier, he had proved that such a code must exist. And so it did. It is part of his legacy; and so is the endless flow of digital information on which the Internet depends, and so is the information omnivory by which we define ourselves as modern.

By his early thirties, he was one of the brightest stars of American science, with the media attention and prestigious awards to prove it. Yet, at the height of his brief fame, when his information theory had become the buzz-phrase to explain everything from geology to politics to music, Shannon published a four-paragraph article kindly urging the rest of the world to vacate his "bandwagon." Impatient with all but the most gifted, he still knew very little of ambition, or ego, or avarice, or any of the other unsightly drivers of accomplishment. His best ideas waited years for publication, and his interest drifted across problems on a private channel of its own. Having completed his pathbreaking work by the age of thirty-two, he might have spent his remaining decades as a scientific celebrity, a public face of innovation: another Bertrand Russell, or Albert Einstein, or Richard Feynman, or Steve Jobs. Instead, he spent them tinkering.

An electronic, maze-solving mouse named Theseus. An Erector Set turtle that walked his house. The first plan for a chess-playing computer, a distant ancestor of IBM's Deep Blue. The first-ever wearable computer. A calculator that operated in Roman numerals, code-named THROBAC ("Thrifty Roman-Numeral Backward-Looking Computer"). A fleet of customized unicycles. Years devoted to the scientific study of juggling.

And, of course, the Ultimate Machine: a box and a switch, which, when flipped on, produced a whirring of gears and a mechanical hand that emerged from the box, flipped the switch off, and disappeared again. Claude Shannon was self-effacing in much the same way. Rarely has a thinker who devoted his life to the study of communication been so uncommunicative. Seen in profile, he almost vanished: a gaunt stick of a man, and a man almost entirely written out of a history defined by self-promoters.

His was a life spent in the pursuit of curious, serious play; he was that rare scientific genius who was just as content rigging up a juggling robot or a flamethrowing trumpet as he was pioneering digital circuits. He worked with levity and played with gravity; he never acknowledged a distinction between the two. His genius lay above all in the quality of the puzzles he set for himself. And the marks of his playful mind—the mind that wondered how a box of electric switches could mimic a brain, and the mind that asked why no one ever decides to say "XFOML RXKHRJFFJUJ"—are imprinted on all of his deepest insights. Maybe it is too much to presume that the character of an age bears some stamp of the character of its founders; but it would be pleasant to think that so much of what is essential to ours was conceived in the spirit of play.

1

1

Gaylord

Here are 110 diamonds, "not one of them small," 18 rubies, 310 emeralds, 21 sapphires, one opal, 200 solid gold rings, 30 solid gold chains, 83 gold crucifixes, five gold censers, 197 gold watches, and one monumental gold punch bowl, and they are exactly where the code said they would be. They are a pirate's hoard, buried five feet down in the South Carolina soil, in the shadow of a gnarled tulip poplar tree. But the tale doesn't end with the treasure; it ends with the code.

William Legrand found it on a parchment washed up from a shipwreck. For months, he sat up learning cryptanalysis by firelight to crack it. And now that the hoard is his, he's content to leave the diamonds counted in a corner while he explains himself at great length to the young man he enlisted to dig it up.

It is simpler than it looks:

> 53‡‡†305))6*;4826)4‡.)4‡);806*;48†8'60))85;]8*:‡*8†83
> (88)5*†;46(;88*96*?;8)*‡(;485);5*†2:*‡(;4956*2(5*-4)8'8*;
> 4069285);)6†8)4‡‡;1(‡9;48081;8:8‡1;48†85;4)485†528806*81
> (‡9;48;(88;4(‡?34;48)4‡;161;:188;‡?;

Count how often the symbols appear and then compare them with the most common letters in the English language. Assume that the most frequent symbol is the most frequent letter: 8 means "E." The most common word in English is "the," so look for a repeated three-letter sequence ending in 8. The sequence ;48 recurs seven times: if it encodes "the," we know that ; means "T" and 4 means "H." Follow those three letters to new letters. ;(88 can only be "tree," and so (means "R." Each symbol solved solves new symbols, and soon the directions to the treasure resolve out of the noise.

Edgar Allan Poe wrote sixty-five stories. This one, "The Gold-Bug," is the only one to end with a lecture on cryptanalysis. It is Claude Shannon's favorite.

Here is where Gaylord, Michigan, ends. The roads turn dirt and give out in potato fields. Main Street is only blocks behind. Ahead are the fields and feedlots, the Michigan apple orchards, the woods of maple, beech, birch, the lumber factory digesting the woods into planks and blocks. Barbed wire runs along the roads and between the pastures, and Claude walks the fences—one half-mile stretch of fence especially.

Claude's stretch is electric. He charged it himself: he hooked up dry-cell batteries at each end, and spliced spare wire into any gaps to run the current unbroken. Insulation was anything at hand: leather straps, glass bottlenecks, corncobs, inner-tube pieces. Keypads at each end—one at his house on North Center Street, the other at his friend's house half a mile away—made it a private barbed-wire telegraph. Even insulated, it is apt to be silenced for months in the ice and snow that accumulate on it, at the knuckle of Michigan's middle finger. But when the fence thaws and Claude patches the wire, and the current runs again from house to house, he can speak again at lightspeed and, best of all, in code.

In the 1920s, when Claude was a boy, some three million farmers talked through networks like these, wherever the phone company found it unprofitable to build. It was America's folk grid. Better networks than Claude's carried voices along the fences, and kitchens and general stores doubled as switchboards. But the most interesting stretch of fence in Gaylord was the one that carried Claude Shannon's information.

Where does a boy like that come from?

Reporting on the wedding of Claude Shannon's parents, the *Otsego County Times* declared itself bamboozled: "Shannon-Wolf Nuptials: Wedding Took Place at Lansing on Wednesday—Date Had Been Kept a Profound Secret." By the paper's account, Claude Shannon Sr. had managed to get married without anyone in town being the wiser.

That Tuesday, August 24, 1909, toward the end of Shannon's third summer in town, a sign appeared on the door of his furniture store: "IF ANYTHING IS WANTED CALL J. LEE MORFORD." That night, Shannon Sr. took the midnight train to Lansing, to the home of the parents of his bride-to-be, Mabel Wolf. "The unconcern which Mr. Shannon displayed as he waited for the train which was nearly an hour late showed that he was perfectly satisfied that no one had gotten wind of his leaving town," reported the paper. The following day, he married Mabel in a quiet ceremony at six o'clock. The bride wore a "wedding gown of white satin with a yoke of lace, and a net veil made with a coronet edged with seed pearls." It seems that the groom concealed the information about the wedding only to keep the party down to a manageable size.

If the paper feigned shock at Shannon's surprise trip to Lansing, the rest of the piece was all small-town sincerity and good wishes. "Mr. Shannon, the groom, has since his residence in this community, made many warm friendships in a business and social way," the paper noted, "and Miss Wolf, the bride, during her many years teaching in the high school here, endeared herself to the people of this community. Mr. and Mrs. Shannon, accept the congratulations of the *Times* and your many friends in this community."

That a run-of-the-mill wedding announcement constituted front-page news says much about the smallness of Gaylord, Michigan. But then, the Shannons were the kinds of people whose wedding date ought to have been common knowledge. Claude Sr. and Mabel were bright threads in Gaylord's fabric. They were neighborly and active in the Methodist church. In downtown Gaylord, two well-known buildings were the work of Claude Sr.: the post office and the furniture showroom with the Masonic lodge tucked upstairs.

Born in 1862 in Oxford, New Jersey, Claude Elwood Shannon Sr. was a traveling salesman who arrived in town just after the turn of the century and bet on its fortunes. He put down his stake—bought out the business dealing in furniture and funerals—and lived to see it pay. "Something which should be found in every home. Nothing more sanitary. The new styles are more attractive. Come in and look over our New Line of furniture," read a typical advertisement in the paper from "C. E. Shannon, The Furniture Man." In Claude Jr.'s childhood, Gaylord was a town of 3,000, and Claude Sr. was a town father: school board, poor board, county fair board, undertaker, Arch Mason, Worthy Patron of the Eastern Star, the kind of Republican for whom the word *staunch* was invented.

His most significant stretch of employment, and the one that earned him the title of "Judge Shannon," was the eleven years he spent as Otsego County probate judge. He settled estates and minor financial disputes, served as notary public, and played the part of local politician and worthy. His service, though modest and conducted in his spare time, was widely appreciated. In 1931, a two-column profile celebrating the twenty-fifth anniversary of his "advent" described Mr. Shannon as "one of our most public-spirited citizens. . . . The years have told the story of a successful business career, due largely to his excellent executive ability and persistency of purpose." Claude Jr., later on, found less to say about him: clever, distant. "He would sometimes help me with my Erector set," he said, "but he really didn't give me much scientific guidance." Claude Sr. was already sixty-nine at Claude's high school graduation; Claude was the son of his old age.

Mabel Wolf was Claude Sr.'s second wife, and she had married him at age twenty-nine, late for a woman of that era. She was eighteen years younger than her husband. Born in Lansing on September 14, 1880, she was a first-generation American. Her father emigrated from Germany to the Union army, survived the Civil War in a sharpshooters' company, and died before he could know Mabel, his last child. Her widowed mother struggled to bring up six children alone in a strange country. Few women in rural Michigan were college graduates—Mabel Wolf was. She arrived in Gaylord with "glowing recommendations" from her professors and took up what was, at the time, the usual work for a woman of intelligence and independence: teaching.

In time, Mabel became principal of Gaylord High, serving in that post for seven years. She was, by all accounts, an active and energetic schoolteacher and administrator. She coached the school's first-ever girls' basketball team and raised money for uniforms and trips. But for all her success, the paper reported the following in 1932:

> At a meeting of the school board it was decided not to hire any married women teachers during the coming school year due to economic conditions. It was decided that when a husband was capable of making a living it would be unfair competition to hire married women. Mrs. Mabel Shannon, Mrs. Lyons, and Mrs. Melvin Cook will be out of the school system due to this ruling.

By that point, at least, there was much in her private life to occupy her. She was a singer and musician of local note. She joined the Library Board and the Pythian Sisters, and she served a term as president of the Gaylord Study Club. When she wasn't volunteering with the Red Cross or the PTA, she lent her contralto voice to town functions and funerals and hosted music clubs in the Shannon living room. In 1905, she landed the leading role of Queen Elizabeth in the operetta *Two Queens* at the local opera house.

Situated in the middle of northern Michigan's central plateau, Gaylord took its name from an employee of the Michigan Central Railroad, which linked many such off-the-beaten-path towns to the rapidly growing hub of Chicago. Gaylord's destiny was shaped by topography and climate perfect for growing millions of acres of forest.

The trees drew the lumber industry, and the first visitors and inhabitants were willing to contend with the climate for the rich cache of white pine and hardwoods. But the environment was austere, with subzero temperatures and thick lake-effect snow. A local history from 1856 concluded, perhaps self-servingly, that the harsh climate offered a brand of moral education: "The fact that [Northern Michigan's] pioneers had more to struggle against in order to provide homes for themselves and the necessary accompaniments of homes developed in them a degree of aggressive energy which has remained as a distinct sectional possession . . . a splendid type of manhood and womanhood—self-reliant, strong, straight-forward, enterprising and moral."

By the time Claude Sr. and Mabel became parents—their daughter, Catherine, was born in 1910, and Claude Jr., the baby of the family, in 1916—the pioneers had come and gone. The town's limits and industries were well established: Gaylord would make itself known for farming and forestry, and a bit of light industry. As the railroads ramified, Gaylord found itself at the intersection of key lines. It became the county seat. Banks and businesses cropped up on Main Street, and the town's population grew and settled around them. But Gaylord remained more village than city, its roots in the making of things: ten pins, sleighs, massive wheels for the transport of timber.

Gaylord was the kind of place in which just about any event was newsworthy. Consider the headlines and snippets from the county newspaper: "WISCONSIN GIRL KILLS WOLF WITH MOP STICK"; "A woman smoking a cigarette on the Midway caused some attention, not all of which was favorable"; "LUMBERJACK DIES OF APOPLEXY"; "VERN MATTS LOSES FINGER"; "MEETING CALLED TO DISCUSS ARTICHOKES." And one September, a paragraph-long ode to a glorious run of fall weather, the lakes like blue mirrors by day and "splotches of silver" by night, a waxing moon bright enough to light up a printed page.

Claude was three years old when the local diner called the Sugar Bowl opened (also headline news). It was, the paper reported, "the first business on Main Street to errect [*sic*] an electric sign outside. Main Street was so dark in those days that the Village Band once gave an after-dark concert under the sign."

Biographies of geniuses often open as stories of overzealous parenting. We think of Beethoven's father, beating his son into the shape of a prodigy. Or John Stuart Mill's father, drilling his son in Greek at the tender age of three. Or Norbert Wiener's father, declaring to the world that he could turn anything, even a broomstick, into a genius with enough time and discipline. "Norbert always felt like that broomstick," a contemporary later remarked.

Compared to those childhoods, Shannon's was ordinary. There was, for instance, no indication in Claude's earliest years of overbearing parental pressure, and if he showed any signs of early precocity, they were not memorable enough to have been written down or noted in the local press. In fact, his older sister was the family's standout: she aced school, mastered piano, and plied her brother with homemade math puzzles. She was also reported to be "one of Gaylord's most popular girls." "She was a model student, which I couldn't quite follow," Shannon admitted. He later suggested that a tincture of sibling rivalry might have driven his initial interest in mathematics: his big sister's talent for numbers inspired him to strive for the same.

Claude had some successes of his own in his early schooling. In 1923, at the age of seven, he won a third-grade Thanksgiving story-writing contest, for his work "A Poor Boy":

> There once was a poor boy who thought that he was not going to have a Thanksgiving dinner for he thought all his playmates would forget him.
>
> Even if they did, one man did not forget him because he thought that he would surprise the little boy early Thanksgiving morning.
>
> So very early on Thanksgiving morning when he awoke, he found a basket of good things at the door. It was filled with so many good things and he was very happy all the day and he never forgot the kind man.

He played the alto horn and performed in the school's musicals. Fifty-nine years later, he still remembered his classmates' names. He wrote to his fourth-grade teacher:

> Some names that come back as through a glass darkly after a half century are Kenny Sisson, Jimmy Nelson, Richard Cork, Lyle Teeter (who committed suicide), Sam Qua, Ray Stoddard, Mary Glasgow, John Kriske, Willard Thomas (a portly boy), Helen Rogers (a portly girl), Kathleen Allen (smart girl), Helen McKinnon (a pretty girl), Mary Fitzpatrick, and of course Rodney Hutchins.

He held in his hands a copy of a black-and-white photo of the fourth-grade class of 1924–25, so reduced in the copying that it took a magnifying glass to resolve the children's faces, and his own eight-year-old face bubbled and then flattened under the moving lens. Gaunt and shy even in those days; piercing eyes. He remembered, too, no doubt from experience, that "boys in those grades tend to fall in adolescent love with their pretty teachers."

Reflecting on his education with the benefit of hindsight, Shannon would say that his interest in mathematics had, besides sibling rivalry, a simple source: it just came easily to him. "I think one tends to get into work that you find easy for yourself," Shannon acknowledged. High school lasted three years for Claude; he graduated a year ahead of the other children in the photo. That said, he wasn't at the top of the class. When a 1932 report in the newspaper recognized three students with straight A's in his high school, Shannon was not among them.

He loved science and disliked facts. Or rather, he disliked the kind of facts that he couldn't bring under a rule and abstract his way out of. Chemistry in particular tested his patience. It "always seems a little dull to me," he wrote his science teacher years after; "too many isolated facts and too few general principles for my taste."

His early gifts were as mechanical as they were mental. Claude's field of vision, for hours at a time, was often the rudder of a model plane or the propeller shaft of a toy boat. Gaylord's broken radios tended to pass through his hands. On April 17, 1930, thirteen-year-old Claude attended a Boy Scout rally and won "first place in the second class wig-wag signalling contest." The object was to speak Morse code with the body, and no scout in the county spoke it as quickly or accurately as Claude. Wig-wag was Morse code by flag: a bright signaling flag (red stands out best against the sky) on a long hickory pole. The mediocre signalers took pauses to think; the best, like Claude, had something of the machine in them. Right meant dot, left meant dash, dots and dashes meant breaks in the imaginary current that meant words; he was a human telegraph.

These gifts were in the family—but perhaps they skipped a generation. It seems that Claude took after his grandfather, David Shannon Jr., the proud owner of U.S. Patent No. 407,130, a series of improvements on the washing machine, complete with a reciprocating plunger and valves for the discharge of "dirt, settlings, and foul matter." David Shannon died in 1910, six years before his grandson's birth. But for a boy of Claude Jr.'s mechanical bent, a certified inventor in the family tree was something to brag about.

And the grandson inherited the tinkering gene. "As a young boy, I built many things, working with mechanical stuff," he recalled. "Erector sets and electrical equipment, built radios, things of that sort. I remember I had a radio controlled boat." One neighbor, Shirley Hutchins Gidden, offered to the *Otsego Herald Times* that Shannon and her brother, Rodney Hutchins, were a conspiratorial pair. "He and my brother were always busy—all harmless projects, but very inventive." She told a different reporter, "Claude was the brains and Rod was the instigator." One experiment stood out: a makeshift elevator built by the two boys inside the Hutchins family barn. Shirley was the "guinea pig," the first to take a ride on the elevator, and it says something about the quality of the boys' handiwork (or her luck) that she lived to tell the tale to a newspaper seven decades later. It was one of many such contraptions, including a trolley in the Hutchins backyard and the private, barbed-wire telegraph. "They were always cooking up something," said Gidden.

Predictably, Claude grew up worshipping Thomas Edison. And yet the affinity between Edison and Claude Shannon was more than happenstance. They shared an ancestor: John Ogden, a Puritan stonemason, who crossed the Atlantic from Lancashire, England, to build gristmills and dams, and with his brother raised the first permanent church in Manhattan, two miles and three centuries from the office where his descendant Claude Shannon would lay the foundations of the Information Age.

It was finished by the spring of 1644, a twin-gabled Gothic church at the island's south tip, hard by the wall of the Dutch fort; the wood shingles on its roof were meant to turn bluish over time and rainstorms, into an imitation of costlier slate. Ogden, who planned it from quarry to weathervane, is said to have been lean, hawk-nosed, and stone stubborn; he was one of the New World's first builders.

Most of us, Claude included, are less demanding than we might be in our choice of idols: from the universe of possible heroes, we single out the ones who already remind us of ourselves. Maybe that's the case for Claude and his distant cousin Edison—even if it was only years after leaving Michigan for good that he discovered the link. Good fortune to learn that one's idol is one's family—and Claude's fortune was better than most.

2

Ann Arbor

A's in math and science and Latin, scattered B's in the rest: the sixteen-year-old high school graduate sent his record off to the University of Michigan, along with an application that was three pages of fill-in-the-blanks, the spelling errors casually crossed out.

> 8. *Have you earned any money during your high-school course?*
>
> Yes.
>
> *How?*
>
> Pedd~~e~~ling papers and del~~e~~ivering telegrams.

The same year he applied to Michigan, his sister graduated from it. Claude was admitted as well, and Ann Arbor was the biggest swarm of humanity he had ever seen.

One hundred and ninety-five miles southeast of Gaylord, Ann Arbor was a city of steep hills and valleys, interrupted by the muddy banks and low gradient of the slow-flowing Huron River. The Huron sealed Ann Arbor's fate as a mill town: sawmills and flour mills punctuated the river banks and powered the economy. Immigrants poured in, most from Germany, but also Greece, Italy, Russia, and Poland. Their ethnic ties ran deep, and churches reinforced the affiliations of caste and clan. By the beginning of the twentieth century, half of Ann Arbor's population was either foreign-born or born to immigrant parents.

It was a population that suffused the city with an irrepressible optimism. On the threshold of a century that would see the Depression and two world wars, a 1901 issue of the *Ann Arbor Argus Democrat* was moved to declare that "the century to come is undoubtedly destined to be the richest and best that man has experienced." After the stock market crashed in October 1929, the *Ann Arbor Daily News* covered the brief recoveries in stock prices rather than report on the devastating declines. Even in December 1929—after more than $30 billion in wealth had evaporated, banks had called in loans, and manufacturing had cratered—Ann Arbor's mayor, Edward Staebler, remained unfailingly buoyant, assuring locals that the economy would recover and that the city would weather the storm.

In the presidential contest of 1932, Ann Arbor defied the state of Michigan. Franklin Roosevelt had won Michigan and forty-one other states in an electoral landslide—but Ann Arbor remained a steadfast Herbert Hoover stronghold. Editorials in the *Daily News* promised recovery and urged voters not to lay the blame for the economy's troubles at President Hoover's feet. His fellow Republicans held on to local offices in Ann Arbor, one of the few places where the president's coattails did more good than harm.

The University of Michigan copied its town's calm confidence. "I am not at all discouraged," university president A. G. Ruthven said in 1931. "I must admit that the curtailment of our resources has permitted me to make certain changes in the organization which I believe will be of lasting benefit." Yet, by the time Claude Shannon arrived at the university in the fall of 1932, that unflinching positivity had run its course. The financial collapse had forced the University of Michigan, Ann Arbor's largest employer and its economic engine, to shave enrollments, halt production on long-planned buildings, and cut pay by 10 percent.

Still, Shannon's timing was fortuitous. Had he arrived a decade or two earlier, he would not have been the beneficiary of the transformation of the university's engineering program during the early years of the twentieth century.

Under the leadership of Dean Mortimer Cooley, an unusually enterprising university administrator, the College of Engineering's "enrollments . . . grew from less than 30 to more than 2,000, the faculty from three instructors teaching several courses to more than 160 professors and staff teaching hundreds of courses, and a temporary shop of 1,720 square feet to over 500,000 square feet of well equipped buildings." The number of engineering students surpassed even the number of students in medicine and law. When it threatened to exceed the enrollment of the university's largest school, the Literary College, Dean Cooley grew excited, and "with his characteristic chuckle, exclaimed [to Professor Harvey Goulding], 'By Jove, Goulding, we'll pass them yet.'" Urbane, well-traveled, and politically savvy, Cooley had first come to the University of Michigan on a Navy billet, as Professor of Steam Engineering and Iron Shipbuilding. Four years later, the Navy allowed him to resign his commission, and the university offered him a proper professorship.

In 1895, the then-dean of the engineering school, Charles Greene, had been asked to create plans for a new building to house the school's growing student body. Greene's request—$50,000 for a small, U-shaped structure—was granted. He died before he could carry out the construction, and Cooley succeeded him as dean. Asked to judge his predecessor's plans and funding needs, Cooley replied, "Gentlemen, if you could but see the other engineering colleges with which we are forced to compete, you would not hesitate for one moment to appropriate a quarter of a million dollars." Something about Cooley's understated certainty swayed the board, and his request was swiftly approved.

A public exhibition in 1913 showcased the spoils of the expansion, as close as a university has probably come to something like a world's fair. Ten thousand people came to tour the facilities and take in the latest technological marvels. Electrical engineers sent messages over a primitive wireless system. Mechanical engineers "surprised their visitors by sawing wood with a piece of paper running at 20,000 revolutions per minute, freezing flowers in liquid air, and showing a bottle supported only by two narrow wires from which a full stream of water flowed—a mystery solved by few." Two full torpedoes, two large cannons, and "a complete electric railway with a block signal system" rounded out the demonstrations. "For the average student as well as for the casual visitor, the Engineering corner of the Campus held mysteries almost as profound as the deeper mysteries of the Medical School," observed one writer.

Cooley's project to expand the engineering college changed the university's core educational program, as well. Eight years before Shannon was born, the college began teaching courses in the theory of wireless telegraphy and telephony, meeting the growing commercial need for engineers trained in wireless transmission. Engineering's rising profile began to draw the attention of deans in other quarters of the university, and disciplinary lines began to blur. By the time Shannon began his dual degrees in mathematics and engineering, a generation later, the two curricula had largely merged into one.

That appealed to Shannon, who admitted that his choice of a dual degree wasn't part of a grand design for his career; it was simply adolescent indecision. "I wasn't really quite sure which I liked best," he recalled. Earning two degrees instead of one wasn't particularly onerous: "It was quite easy to do because so much of the curriculum was overlapping. I think you needed two extra courses and some summer school to get degrees in both fields," said Shannon. Those studies gave him his first taste of communication engineering, which he found "especially to my liking" for its blend of practice and theory—because it was "the most mathematical, I would say, of the engineering sciences."

Though the dual degree was common enough, Shannon's variety of indecision, which he never entirely outgrew, would prove crucial to his later work. Someone content to build things might have been happy with a single degree in engineering; someone drawn more to theory might have been satisfied with studying math alone. Shannon, mathematically *and* mechanically inclined, could not make up his mind, but the result left him trained in two fields that would prove essential to his later successes.

He joined Radio Club, Math Club, even the gymnastics team. Shannon's records of leadership during this time are two. One is his stint as secretary of the Math Club. "A feature of all meetings," a journal recorded, "was a list of mathematical problems placed on the board and discussed informally after the regular program. A demonstration of mathematical instruments in the department's collection made an interesting program." The other was news enough that the hometown paper saw fit to print it as an item of note: "Claude Shannon has been made a non-commissioned officer in the Reserve Officers Training Corps at the University of Michigan."

In the Engineering Buildings, where Claude spent the bulk of his time, his classmates tried the strength of shatterproof windshield glass, worked to muffle milk-skimming machines, floated model battleships on a sunless indoor model sea. But the real life on campus was outside the classroom.

In the spring of 1934, Claude's sophomore year, an unusually misanthropic editor got a hold of the yearbook's anonymous comedy section and turned it into an account of student life narrated by an escaped mental patient convinced he's an anthropologist:

> Breakfast at the dining hall: "The stories of last week-ends parties assume a fundamental sameness . . . 'We went to ______'s (dance hall, night club, apartment, or fraternity) and had______ highballs,______beers and______shots of______. After the party______got sick and______and I had to carry him all the way from______to______.'"

Someone spills his glass of orange juice on a coed's lap, and everyone laughs for five minutes, until they forget what they're laughing about and go silent again. "It is very quiet now. . . . The business of laughing seems to have taken something out of every one." Breakfast breaks up at eleven and they spend the rest of the morning going through the motions of hilarity.

Yearbook blurbs on the big men on campus were usually a string of mild in-jokes, but there was some acid to them in the spring of 1934. There is the track star who each night "removes his legs (they being in some wise attached to his body) and places them in a gold and glass case for all and sundry to admire." The student politician, "parading down State street with seven stooges at his heels, well fortified from contradiction or blasphemy." The newspaper editor, "wistfully pounding a typewriter in the secrecy of his cupboard office, attempting to veil the fact that he has nothing to veil."

Claude, by contrast, was a small man on campus. But he and the editor may have had a hunch in common: the introverted suspicion that they are surrounded by animate machines, detachable parts and all, all surface and funny motions. It takes a cynic or an engineer to discover "the business of laughing." Later on, a girlfriend remembered Claude's own laugh: "He laughed in small explosions as though he were coughing, and had never quite learned how to be merry." It was his own funny motion of diaphragm and throat.

In the spring of Shannon's sophomore year, a stroke ended his father's life. For fifteen months Claude Sr. had fought illnesses and lived confined at home, his seventy-one years catching up with him. In the days after his death, the town of Gaylord shut down in his honor. The funeral was at the Shannon home at two o'clock on a Tuesday afternoon; the pallbearers, Claude Sr.'s business associates, were august. By Wednesday, Claude was on his way back to the university.

Soon after his father's death, something broke between Claude and his mother. His sister was grown and gone, the town father was in the ground, and Claude and Mabel were alone together for the first time. It ended disastrously. It seems the rupture was caused, absurdly enough, by a plate of cookies: she saved the good cookies for guests and offered Claude only the burned ones. Whatever the cause, Claude spent his remaining school vacations at an uncle's. He and his mother would barely interact for the rest of his life.

He completed his time as a student, distinguishing himself enough to earn admission as a senior to both the Phi Kappa Phi and Sigma Xi honor societies. In the spring of 1934, at the age of seventeen, Claude Shannon claimed his first publication credit, on page 191 of the *American Mathematical Monthly*. He had worked out the solution to a math puzzle and landed a spot in the "Problems and Solutions" section. The editors of the section welcomed "problems believed to be new, and demanding no tools beyond those ordinarily furnished in the first two years of college mathematics." The problem Shannon solved had appeared in the previous fall:

E 58 [1933, 491]. *Proposed by R. M. Sutton, Haverford College, Pa.*

In the following division of a three-place number into a five-place number each digit has been replaced by a code letter. Assuming only that the remainder, *Y*, is not zero, reconstruct the problem and show that the solution is unique.

```
L M N)R S T U N(U X
      R T Y X
      -------
        T Y Y N
        T Y Y J
        -------
              Y
```

Coming as it did in the back of the journal, after the weightier work of math papers and book reviews, Shannon's six-part solution to the problem was nothing notable—except for the fact that it existed at all, a sign that his childhood fascination with codebreaking was starting to pay adult dividends. Buoyed, we imagine, by this first success, Shannon again submitted a solution and was again published in the *Monthly*'s back pages, in January 1935, in answer to this problem:

E 100 [1934, 390]. *Proposed by G. R. Livingston, State Teachers College, San Diego, California.*

In two concentric circles, locate parallel chords in the outer circle which are tangent to the inner circle, by the use of compasses only, finding the ends of the chords and their points of tangency.

Modest as they are, these early efforts are a window into the education of Claude Shannon. We can infer from them that the college-aged Shannon understood the value of appearing in a professional public forum, one that would earn the scrutiny of mathematicians his age and the attention of those older than him. That he was reading such a journal at all hints at more than the usual attention paid to academic matters; that his solutions were selected points to more than the usual talent. Above all, his first publications tell us something about his growing ambition: taking time out from the usual burdens of classes and college life to study these problems, work out the answers, and prepare them for publication suggests that he already envisioned something other for himself than the family furniture business.

His something other would begin, in earnest, with a typed post-card tacked to an engineering bulletin board. It was an invitation to come east and help build a mechanical brain. Shannon noticed it in the spring of 1936, just as he was considering what was to come after his undergraduate days were over. The job—master's student and assistant on the differential analyzer at the Massachusetts Institute of Technology—was tailor-made for a young man who could find equal joy in equations and construction, thinking and building. "I pushed hard for that job and got it. That was one of the luckiest things of my life," Shannon said later. Luck may have played a role, but the application's acceptance was also a testament to the keen eye of a figure who would shape the rest of Shannon's life and the course of American science: Vannevar Bush.

3

The Room-Sized Brain

If you were searching for the origins of modern computing, you could do worse than to start here: on Walnut Hill, just north and west of Boston, in 1912, where an overdressed lawnmower man was trudging up a grassy incline behind his machine. He took a moment to pose for a grainy photo, hands on the tiller, eyes on his work, face turned from us; the white of the grass, the black of his two-piece suit, the black of the machine. You'd deduce in a second, of course, that its purpose is something stranger than lawn care: the tall grass is untouched, and where there ought to be blades there is a blank box, riding slung between two bicycle wheels.

It was the failed first invention of a college senior, and though it ran just as promised, it bored nearly everyone beyond its twenty-two-year-old creator. Inside the box hung a pendulum, and a disc powered by the back bicycle wheel. Resting on the disc were two rollers: one measured vertical distance and wielded a pen, one measured horizontal distance and turned the drum of paper beneath. It was a geography machine, a device aimed to put land surveying teams out of business. Using the old method, heavy on telescopes and trigonometry, three men could cover three miles of ground per day, and at day's end, they'd have tables of data to convert into a cross-section picture of the land over which they'd slogged. The college senior claimed that he, working alone, could nearly triple their speed—and he did it by skipping straight to the picture. Inside the body of his Profile Tracer was the lay of the land spooled on a rolling drum, drawn in ink by a machine so accurate that if "it ran over a manhole cover, it would duly plot the little bump."

It earned a patent, and simultaneous bachelor's and master's degrees for its creator, but little else. He made the corporate rounds and failed to sell a single one, or even the license for the patent—his cold letters unanswered, his pitch meetings over in minutes. And even if he could have said, in a hypothetical moment of awesome clairvoyance, "Look, in twenty years the guts of this lawnmower will run the most powerful thinking machine that human hands have ever built"—it would have sounded close to gibberish. But it would also have been true.

The man in the black suit is Vannevar Bush, and this photo marks his start. Pugnacious and perpetually time-strapped, grandson and great-grandson of Yankee whaling captains, saddled with a name so frustratingly hard to pronounce that he would instruct others to call him "Van" or even "John"—the twenty-two-year-old inventor would one day be, although he couldn't possibly imagine it yet, the most powerful scientist in America.

He would preside over a custom-made brain the size of a room. He'd counsel presidents. He'd direct the nation's scientists through World War II with the same brusqueness with which he once imagined unemploying two-thirds of the surveying profession. *Collier's* magazine would call him "the man who may win or lose the war"; *Time*, "the general of physics."

And not least among these accomplishments would be this: he'd be the first to see Claude Shannon for who he was.

"Suppose," said Vannevar Bush—two decades older, now a doctor of engineering and vice president of MIT—"an apple drops from a tree." It's just as well that he started with an example from the high school physics chalkboard. Mathematically speaking, he was a man of only moderate brain, "fourth or fifth echelon" by his own admission.

But he was blessed with brilliant hands. He'd been—like Claude Shannon, his greatest student—a basement tinkerer from his earliest memories. Much of his adult life was spent, it turned out, building dogged, untiring mathematical brains of wood and metal, brains that in some ways far outclassed his own—and that would ultimately be the scene of Shannon's first breakthrough.

"The thing we know about that apple," Vannevar Bush continued, "is, to a first approximation, that its acceleration is constant." We can plot its fall on the chalkboard in seconds. "But suppose we want to include the resistance that air offers to the fall. This just puts another term in our equation but makes it hard to solve formally. We can still very readily solve it on a machine. We simply connect together elements, electrical or mechanical gadgets, that represent the terms of the equation, and watch it perform."

What is it about an apple in a physicist's vacuum that needs only pencil and paper, while an apple falling through the air of the real world demands solution by gadget? Both falls, as Bush noted, can be captured in differential equations—the equations at the heart of calculus that represent continuous change. So first imagine the apple falling on the head of, say, Isaac Newton (and it's no coincidence that the man who formulated the laws of gravitation also co-invented calculus—without equations that capture change over time, there's no making sense of gravity). In a vacuum, the apple falls 9.8 meters per second faster, each second, until it concusses Newton.

But now drop the apple on Newton in the open air. Gravity's force, of course, doesn't change. But the faster the apple falls, the greater the resistance of the air pushing back against it. The apple's acceleration now depends on both the gravity speeding it up and the air resistance slowing it down, which in turn depends on the apple's speed at any moment, which in turn is changing every fraction of a second. *That* is the kind of problem that calls for a more-than-ordinary brain.

How fast can a population of animals grow before it crashes? How long before a heap of radioactive uranium decays? How far does a magnet's force extend? How much does a massive sun curve time and space? To ask any of these questions is to ask for the solution to a differential equation.

Or, of special interest to Bush and his electrical engineering colleagues: How great a power surge could the nation's electrical grids tolerate before they failed? Given all the wealth and work it had taken to electrify America, it was a multimillion-dollar question. In the 1920s, reflected one of Bush's graduate students, transmitting power from state to state was like "the towing of one car by another with a long elastic cable stretched almost to the breaking point. Under these conditions, any mishap, such as a short circuit or a sudden adding of load, would in effect snap the towing cable." By 1926, engineers had discovered the equations that could predict the cable's snapping point. The catch was that solving these power equations meant a long and error-prone slog: doing calculus by hand, graphing the results by hand, finding the area covered by the graphs by tracing their outline with a rolling mathematician's tool called a planimeter, and then inserting the area figures into further equations—all of which meant that the lights would flicker and die long before the work was done.

It turned out that most differential equations of the *useful* kind—the apple-falling-in-the-real-world kind, not the apple-falling-down-a-chalkboard kind—presented just the same impassable problem. These were not equations that could be solved by formulas or shortcuts, only by trial and error, or intuition, or luck. To solve them reliably—to bring the force of calculus to bear on the industrial problems of power transmission or telephone networks, or on the advanced physics problems of cosmic rays and subatomic particles—demanded an intelligence of another order.

By the time Bush and his students set to work, scientists had been after such a brain for two generations. Long before it was needed to stabilize the electrical grid, it was sought for a much more ancient problem: predicting the ocean tides. For sailors, tide knowledge dictated when to come into harbor, where to fish, and even when to launch invasions. If little fishing boats could rely on guesswork and memory, the iron-sided, steam-belching ships of the nineteenth century required something more precise. And there was no precision to be had in simply looking at high-tide marks and waiting for the sea to repeat itself, because the simple model of Newton's airless world—moon and sun tugging on the water at each day's appointed times—fell into seeming chaos when confronted with the reality of each shoreline's special shape and each seabed's unseen slope. From the God's-eye view, there is a law of tides; from our earthbound view, only some petty local ordinances.

But a half century after Newton, mathematicians found that the most chaotic-seeming fluctuations—from stock prices to tide charts—could be broken down and represented as the sum of much simpler functions, wavelike patterns that did indeed repeat themselves. Anarchy concealed order; or rather, anarchy was dozens of kinds of order happening at once, all shouting to be heard over one another. So how to find the order in the tides?

In 1876, the wizard-bearded Scots-Irish physicist William Thomson—later ennobled as Lord Kelvin, a name he took from the river that flowed by his laboratory—proposed to do it by machine. At Thomson's Cambridge graduation exam, the professor questioning him leaned over to his colleague and whispered, "You and I are just about fit to mend his pens." From his days in school, he'd kept as a personal motto some lines from Alexander Pope: "Go, wondrous creature! mount where Science guides; / Go measure earth, weigh air, and state the tides." And while the poet surely meant to speak to Man in the aggregate, Thomson could hardly be blamed if he ever imagined that he himself was the creature addressed.

Thomson's tidal solution was something like the inverse of Bush's lawnmower. The surveying machine would read the land's data of hills and dips and even manhole covers and output a graph; the tide machine invented by Thomson and his brother, which they christened the harmonic analyzer, took a graph as input. The operator stood before a long, open wooden box resting on eight legs, a steel pointer and a hand crank protruding from its innards. With his right hand, he took hold of the pointer and traced a graph of water levels, months' data on high tides and low; with his left, he steadily turned the crank that turned the oiled gears in the casket. Inside, eleven little cranks rotated at their own speeds, each isolating one of the simple functions that added up to the chaotic tide. At the end, their gauges displayed eleven little numbers—the average water level, the pull of the moon, the pull of the sun, and so on—that together filled in the equation to state the tides. All of it, in principle, could be ground out by human hands on a notepad—but, said Thomson, this was "calculation of so methodical a kind that a machine ought to be found to do it."

And so it had been. With the equation extracted from the surf, a tide table was no longer just a record of the past, but a promise of the future. Draw the table as a graph; feed the graph into the harmonic analyzer; and finally, use the analyzer's readings to custom-rig Thomson's next invention, a fifteen-pulley mechanical calculator the size of a wardrobe that drew, with pen and ink, its own graph of tide levels for the year to come. In 1876, the tide predictor could accurately draw a year's worth of the future in four hours; by 1881, it took twenty-five minutes.

It was politely received and politely shunted aside. Even in 1881, few practical equations were susceptible to mechanical solution, so it seemed wiser to go on paying pencil-pushers than to mass-produce a device with such a limited scope. Perhaps, too, Thomson's fellow mathematicians took offense at the thought that any part of their work could be automated as easily as the labor of a factory hand. Most important, though Thomson conceived of a truly versatile problem-solving machine, the crucial component was effectively missing until world war brought a new impetus to the search.

So imagine, now, not a ship coming into harbor with the tide, but a dreadnought rolling on a choppy sea, readying its guns to lob an explosive shell at a moving target more than ten miles over the horizon. Imagine a sea battle between two floating arsenals that would remain, until the very end, mutually invisible. At that distance, the pitch of the waves, the density of the air at each level of the projectile's trajectory, the curvature of the earth, even the earth's rotation during the time of the shell's flight would all conspire to determine whether that shell would hit water or steel. Each of those factors formed a variable in—again—a differential equation. A naval battle at that range was not simply a gunfight, but a mathematical race (in which the reward for second place was often a watery grave). In the First World War's largest naval engagement, the Battle of Jutland, in 1916, every British ship but one steamed into battle with human-directed guns. They hit only 3 percent of their targets; they lost more than 6,000 men. With those stakes, a reliable equation-solving machine was suddenly worth the cost.

It was Hannibal Ford, a mechanical engineer from upstate New York, who supplied Thomson's missing part. He'd gotten his start taking apart watches and clocks, and then moved on to work on typewriters. Where Thomson chose as his college watchword an heroic couplet from Pope, Ford's page in the Cornell yearbook had an earthier motto: "I would construct a machine to do any old thing in any old way." The machine he had constructed by 1917 automated a key step in the solution of differential equations: it found integrals, or the area under curves (including the curve of a shell in flight). Long before electronics, it could all be done mechanically. In the case of Ford's integrator—nicknamed the "Baby Ford" by grateful American sailors—two ball bearings rested on the surface of a flat, spinning disc. They were free to move continuously across the disc's surface: the farther from the center, the faster they'd spin. The distance from the center stood for the shape of the equation's curve, and the speed of their spinning stood for the answer. The ball bearings turned a cylinder that powered the rest of the machine and transmitted the answer, through gears and gauges, to the gunners. Given inputs including the speed and course of the attacking ship and the enemy ship, the Baby Ford would generate the range to the target, the direction for fire, and the time the shell would be in the air. These, in turn, would dictate the angle of the guns.

Hannibal Ford was not the first to imagine such a machine. But his machine was among the first to find integrals reliably, let alone below-decks on a ship tossed by waves and shaken by exploding shells, when a slip of a ball bearing from its orbit would send the crew back to the days of spyglasses and intuition. It was, said Vannevar Bush, "a marvel of precision and completeness." Soon Bush would run six of them at once—and he'd set them to search not just for the pitch of a gun, but for the shapes of atoms and the structures of suns.

Thomson's harmonic analyzer, Ford's integrator, Bush's Profile Tracer: conceived in isolation from one another, single-purpose machines built to answer only one specialized question apiece, they still had a crucial quality in common. They were all working models of the physical world—of the slope of a hill or the fall of a shell—simplified down to the essence. They were all, in a way, bare-bones miniatures of the processes they described; they were, in other words, resolutely analog. But it was Vannevar Bush who brought analog computing to its highest level, a machine for all purposes, a landmark on the way from tool to brain. And it was Claude Shannon who, in a genius accident, helped obsolete it.

Bush would later recognize his computer's precedents in Thomson and Ford. But when he first set to work in the mid-1920s, searching for a way to shrink America's power network to the size of his lab, he was largely ignorant of his analog forebears. Where, then, did he start?

In a sense, he started as a teacher. As well as an inventor, Bush was an instructor of young engineers at a time when MIT's electrical engineering department was coming into national prominence. Fall in Cambridge, Massachusetts, would begin with an auditorium full of bright freshmen, slacks pressed and hair freshly combed, sitting stunned as Bush punctured their self-regard. He would rise at the lectern, hold up a simple pipe wrench, and offer a simple challenge: "Describe this."

One by one the freshmen would take their shot, and one by one their descriptions were dismantled: Bush would show how each definition was so vague that it could apply to any number of wrenches, not *this* wrench in front of them. And he would conclude by reading out the exact and correct patent application:

> By turning the nut to the right or left the movable jaw may be moved either toward or away from the fixed jaw, as may be desirable. The inner face of the movable jaw is formed at a right angle to its shank, and is also provided with a series of teeth, which pitch or rake on its fellow jaw. . . . The sliding or movable jaw [may] be projected outward so as to stand at an outward inclination with respect to the other jaw, in order to enable the jaws to be readily applied to a pipe. . . .

And so on.

The point was precision. In particular, the point was rigor in reducing the hard, solid world—the wrench—into symbols so exact—the patent application—that one could be flawlessly translated from the other. Given the pipe wrench, produce the words for that wrench and no other; given the words, produce the wrench. That, Bush taught his students, was the beginning of engineering.

For the same reason—rigor in symbolizing the world—every engineer was taught to draw. Leave pure numbers for pure mathematicians—engineers would learn math with their hands. "A man learns to use the Calculus as he learns to use the chisel or the file," said one reformer who helped give engineering education its practical bent in the early century. A math laboratory of that era was "well-stocked with clay, cardboard, wire, wooden, metal and other models and materials"—and with graph paper, which was only about as old as Bush was. At Bush's MIT, math and engineering were an extension of the metal shop and the woodshop, and students who were skilled with the planimeter and the slide rule had to be skilled as well with the soldering iron and the saw. There is perhaps a source here for engineers' persistent status anxiety, "uncertain always where they fit," as the great critic Paul Fussell put it, "whether with boss or worker, management or labor, the world of headwork or the world of handwork." But there was also the conviction that handwork *was* headwork, as long as the translations had precision. Given precision, an equation could be grasped and solved in pictures and motion, just as a wrench could be pinned down by the right words.

Working with a mechanic to build his early analog computers, Bush came to see just how thoroughly calculus could be learned by hand: "He had learned the calculus in mechanical terms," Bush explained, "a strange approach, and yet he understood it. That is, he did not understand it in any formal sense, but he understood the fundamentals; he had it under his skin."

In the whirring of their integrators and the turning of their gears, Bush's machines *embodied* calculus. Like good engineers, they took drawings as input and gave drawings as output. They might have happened anywhere—but it's hardly surprising that they were pieced together in an engineering department.

By 1924, Bush and his students had built an integrating machine that improved on Ford's. By 1928, in search of the solution to a stable grid, they were able to model 200 miles of power lines in a fifty-square-foot lab. The same year, work started on an all-purpose analog computer: the differential analyzer. When it was finished, three years and $25,000 later, it was a brain the size of a room, a metal calculus machine that could whir away at a problem for days and nights on end before it ground to a halt. One problem, which measured the effects of the earth's magnetic field on cosmic rays, took thirty weeks of spinning gears—but when it was done, the differential analyzer had solved, by brute force, equations so complex that even trying to attack them with human brainpower would have been pointless. Indeed, Bush's lab now owned the computational power to turn from the problems of industry to some of the fundamental questions of physics.

"It was a fearsome thing of shafts, gears, strings, and wheels rolling on disks," said an MIT physicist who turned to the differential analyzer to study the behavior of scattering electrons, "but it worked." It was an enormous wooden frame latticed with spinning rods, resembling a giant's 100-ton foosball set. At the input end were six draftsman's tables, where the machine read the equations it was to evaluate, much like Thomson's analyzer read a graph of the tides. The operators turned hand cranks that sent the machine's pointers over a hand-drawn graph of the equation to be analyzed: "for example," read one contemporary account, "in calculating the scattering of electrons by an atom, it is necessary to supply the machine with the relation between the potential of the atomic field and the distance from the centre of the atom." In this way, the details of the equation were communicated to the machine's internal shafts. Each shaft represented a variable (the current in a power line, or the size of an atomic nucleus); the greater the variable, the faster the shaft spun. These, in turn, drove integrators like Ford's: A flat disc spun in place, and standing perpendicular on the disc was an integrating wheel. The farther the operators had placed the wheel from the center of the disc, the faster it turned. The wheel was linked to five more integrators of identical construction. At the very end, the speed of the integrating wheels drove a pencil that moved up and down as the graph paper underneath it unwound at a continuous rate. The question was a graph, and at the end, after days or even months of revolutions, the answering graph appeared.

The mathematics were infinitely more complex—but Vannevar Bush's lawnmower might have recognized in this calculating room a distant descendant. The differential analyzer, wrote one science historian, "still interpreted mathematics in terms of mechanical rotations, still depended on expertly machined wheel-and-disc integrators, and still drew its answers as curves. Differential equations and contours of elevation—Bush's computers were very much the offspring of the early Profile Tracer."

This was the computer before the digital revolution: a machine that literally performed equations in the process of solving them. As long as the machine was acting out the equations that shape an atom, it was, in a meaningful sense, a giant atom; as long as it was acting out the equations that fuel a star, it was a miniature star. "It is an analogue machine," said Bush. "When one has a problem before him, say the problem of how a bridge that has not been built will sway in a gusty wind, he proceeds to make a combination of mechanical or electrical elements which will act in exactly the same manner as the bridge—that is, will obey the same differential equations." For the physicist or engineer, two systems that obey the same equations have a kind of identity—or at least an analogy. And that, after all, is all our word *analog* means. A digital watch is nothing like the sun; an analog watch is the memory of a shadow's circuit around a dial.

The computer clacked and hummed and scribbled away, spinning out its analogies, and when it ran through the night, Bush's students kept watch by its side in shifts, ears tuned for the sound of a wheel slipping its orbit. On the nights when it all ran smoothly, they struggled to stay awake in the humming room. And so passed five years.

4

MIT

Claude Shannon was, at least, acquainted with the cold. The wind that blew off the Atlantic was saltier than Michigan's but no more chilling; the New England snow was almost as deep. Twenty years old, displaced from the Midwest for the first time, and alone, surely he would have taken the familiar where he could find it. For those unable to bear the cold, though, MIT offered corridors and tunnels, long expanses painted in Institutional Gray. The engineers could spend an entire winter indoors. They could virtually live in the gray tunnels, and there were many days that Shannon never saw sun—with the special exception of "MIThenge," the two winter days each year when the sun aligned with the axis of the corridors, and lit the gray walls gold as it set.

"Institute folklore has it that an alert eye can sometimes pick out pencil lines on the corridor walls, shoulder high and parallel to the floor," writes MIT historian Fred Hapgood. "These are supposed to be the spoor of members of the community so adapted to the corridors that they can navigate them blind . . . press a pencil against the wall, roll their eyeballs up into their heads, go out-of-body after some complex problem, and wheel away on automatic pilot." On greener days, a walk outside would take Shannon past columned facades chiseled with the names of the greats: Archimedes, Copernicus, Newton, Darwin.

MIT was a neoclassical island in what was still an industrial Boston suburb, and the Pantheon-style dome at its center sat as an uneasy neighbor to the factories and mills along the Charles River. The dome sitting atop tunnels was itself a compromise between architects: one who wanted the new campus to bear at least some comparison to that other college up the river, and another who insisted that it be built on the principles of "efficiency and *avoidance of lost motion* by student and teacher, equal to that which obtains in our best industrial works." This was MIT's place in the world, in miniature: an adjunct to industry with aspirations to "purer" science—both factory and dome.

The buildings themselves were a tribute to the quantitative mindset, known more by number than by name. It was a postcard about Bush's analyzer that brought Shannon to Building 13, and it was Bush who approved his application and offered him admission to the graduate program. Both were engineers in a hurry. The better to get on with the work of earning money and supporting a family, Bush had managed simultaneous undergraduate and graduate degrees; Shannon had finished high school in three years, two BS degrees in four, and now was on to graduate work with barely a summer's pause. It was a sign of Bush's regard for his new pupil that he placed him in charge of the analyzer's most advanced and most finicky part.

By 1935, a year before Shannon arrived in Cambridge, the differential analyzer had reached its limits. Mechanical contraption that it was, with each new equation it had to be deconstructed and reassembled again. What Bush and his team had built was not so much a single machine as a huge array of machines in outline, to be rebuilt for every problem and broken down at every solution. It was versatility at the cost of efficiency; and because the analyzer's entire mission was to bring efficiency to computations that could at least in theory be worked out by human hands, these recurring bottlenecks compromised its whole reason for being.

In response, Bush dreamed of an analyzer that could essentially reassemble itself on the fly: one with automatic controls enabling it to turn from equation to equation without pausing, or even to solve multiple interacting equations simultaneously. Switches would take over the work of screwdrivers. At a time when Bush's ambitions far exceeded MIT's Depression-era budget, he was still able to secure $265,000 from the private philanthropists at the Rockefeller Foundation to develop this next-generation computer. And he brought Claude Shannon to MIT to help him run it.

So for the next three years, Shannon's world was gray corridors and the walls of the humming room: and within that room, the walls of a little box of 100 switches closing and opening, fastened to the analyzer, a world within a world. In the box were the brains of the brain, the switches and relays that now controlled the machine and rebuilt it as it spun, each relay, writes James Gleick, "an electrical switch controlled by electricity (a looping idea)." Open. Close. Weeks and months of this.

What happened when Claude Shannon flipped a switch? Think of a switch or relay as a drawbridge for an electrical current: closed, the switch would allow the current to pass on to its destination; open, the switch would stop the current in its tracks. The destination might be another relay, which would then open or close on the basis of the input it received—or it might, in the easiest example, be something as simple as a little electric light. All of this was second nature to Shannon, from as far back as the Gaylord Western Union office and his barbed-wire network. And it was systematized for him in Ann Arbor, where he dutifully drew circuit diagrams along with the rest of the electrical engineers. Series: the current has to pass through both of two switches before it illuminates the light; parallel: the current is free to pass through either, or both.

These were the blocks that comprised the hundred-switch logic box attached to the differential analyzer, or the electric guts of assembly lines, or the million-relay system that controlled the nation's telephone network. There were circuits built to transmit a current when two switches were closed, but not zero or one or three; there were circuits drawn as branching trees or symmetrical deltas or dense meshes, an entire electric geometry that Shannon had learned by heart. And, in the old tradition of engineers, it was all rigged by hand, drawn step by step on blackboards or just soldered together in the belly of the machine, the only proof of the circuit's rightness in the tangible results: whether the call went through, whether the wheel spun edgewise on its disc, whether the light lit. Circuits before Shannon were like differential equations before the analog computer: errors for every trial until the errors stopped, and nothing any cleaner. Building circuits in those days was an art, with all the mess and false-starting and indefinable intuition that "art" implies.

But here was Shannon shut in a room with a machine built to automate thought—built, in the name of industry and efficiency, to remove the art from math. And in the midst of his work, he came to understand that he knew another way of automating thought, one that would ultimately prove far more powerful than the analog machine.

How is logic like a machine? Here is how one logician explained it around the turn of the twentieth century: "As a material machine is an instrument for economising the exertion of force, so a symbolic calculus is an instrument for economising the exertion of intelligence." Logic, just like a machine, was a tool for democratizing force: built with enough precision and skill, it could multiply the power of the gifted and the average alike.

In the 1930s, there were only a handful of people in the world who were skilled in both "symbolic calculus," or rigorous mathematical logic, and the design of electric circuits. This is less remarkable than it sounds: before the two fields melded in Shannon's brain, it was hardly thought that they had anything in common. It was one thing to compare logic to a machine—it was another entirely to show that machines could *do* logic.

In Michigan, Shannon had learned (in a philosophy class, no less) that any statement of logic could be captured in symbols and equations—and that these equations could be solved with a series of simple, math-like rules. You might prove a statement true or false without ever understanding what it *meant*. You would be less distracted, in fact, if you chose not to understand it: deduction could be automated. The pivotal figure in this translation from the vagaries of words to the sharpness of math was a nineteenth-century genius named George Boole, a self-taught English mathematician whose cobbler father couldn't afford to keep him in school beyond the age of sixteen. Not long before Thomson conceived of his first analyzer, Boole had proven himself a prodigy with a book that fully earned its presumptuous title: *The Laws of Thought*. Those laws, Boole showed, are founded on just a few fundamental operations: for instance, AND, OR, NOT, and IF.

Say we'd like to designate all the people in London who are blue-eyed and left-handed. Call the property of blue eyes x, and call the property of left-handedness y. Use multiplication to stand for AND, addition to stand for OR, and a simple apostrophe (in lieu of a minus sign) to stand for NOT. Remember that the goal of all this is to prove statements true or false; so let 1 stand for "true," and 0 for "false." Those are all the rudiments for turning logic into math.

So the set of all Londoners who are *both* blue-eyed and left-handed simply becomes xy. And the set of all Londoners who are *either* blue-eyed or left-handed becomes $x + y$. Imagine, then, that we want to evaluate the truth of the statement, "This particular Londoner is a blue-eyed left-hander." Its truth depends on the truth of x and y. And Boole set out the precepts for assigning the statement a 1 or a 0 given what we know about x and y:

$$0 \cdot 0 = 0$$
$$0 \cdot 1 = 0$$
$$1 \cdot 0 = 0$$
$$1 \cdot 1 = 1$$

It's easy to translate those equations back into English. If the Londoner is neither blue-eyed nor left-handed, the statement we're trying to evaluate is of course false. If the Londoner is only blue-eyed or only left-handed, the statement is again false. Only if the Londoner is both does the statement become true. In other words, the operator AND only gives "true" if all of the propositions it's operating on are true.

But Boolean algebra is more than a rehash of ordinary math. Imagine now that we want to evaluate the statement, "This particular Londoner is blue-eyed *or* left-handed." In that case, we get the following:

$$0 + 0 = 0$$
$$0 + 1 = 1$$
$$1 + 0 = 1$$
$$1 + 1 = 1$$

If the Londoner is neither blue-eyed nor left-handed, the statement is false. But if he is blue-eyed, or left-handed, or both, it's true—so in Boole's algebra, 1 + 1 equals 1. The operator OR gives true if any proposition it's operating on is true, or if all are. (Boole also recognized another kind of "or," called the exclusive-or, which only gives true if one or the other proposition is true, but not both.)

From these simple elements (as simple as switches, Shannon reflected) we can build our way to progressively more complicated results. For instance, we can prove that $x + xy = x$, or that the truth-value of the statement "either x, or both x and y" depends solely on the truth of x. Or we can prove that $(x + y)' = x'y'$: in other words, "x or y" is false when "both not-x and not-y" is true, and vice versa. This, Boole argued, is all there is of logic. *X* and *y*, and as many other variables as we choose, can stand for whatever statements we want, as long as they're either true or false—and with the simple, nearly mindless operation of a few rules, we can deduce from them everything that can be deduced. A mechanical logic means no more puzzling over "All men are mortal, Socrates is a man . . ." and so on: nothing but symbols, operations, and rules. It took a genius to lay down the rules, but it would only take a child to apply them. Or something simpler than a child.

It was all very interesting, but for nearly a century little that was practical came of it. It was taught to generations of students mainly as a philosopher's curiosity, and so it was taught to Claude Shannon. At the time, he said, he mainly enjoyed it for the sound of the word: "Booooooolean." But something of it stayed with him as he struggled to make sense of the hundred-switch box; something of Boole's simplicity stood out alongside the devilishly complex equations he was solving for Bush. Close, open. Yes, no. 1, 0.

Something of it stayed with him, too, as he made his way from

MIT to New York for the summer of 1937. If one other group of people in the world was edging closer to thinking simultaneously about logic and circuits, it was the minds at Bell Laboratories, who had taken Shannon on for a summer internship. Only a temporary hire, Shannon was likely occupied with the ordinary duties assigned to interim help, and his 1937 summer left no marks in the Labs' records. But he brought to the Labs a deep sense of mathematical logic and a better-than-average knowledge of circuit design—and the nagging feeling that the two were connected. He brought them, moreover, to the heart of the phone company, the owner of the most complicated, far-reaching network of circuits in existence, and his work was part of the mathematical effort to make that network perform better and cost less.

Crucially, it was around this time that he first put pen to paper and began tying together the commonalities he sensed in Bush's analyzer, Bell's network, and Boole's logic. Half a century later, Shannon tried to recall his moment of insight, and tried to explain how he could have been the first to understand what the switches meant. He told a journalist,

> It's not so much that a thing is "open" or "closed," the "yes" or "no" that you mentioned. The real point is that two things in series are described by the word "and" in logic, so you would say this "and" this, while two things in parallel are described by the word "or." . . . There are contacts which close when you operate the relay, and there are other contacts which open, so the word "not" is related to that aspect of relays. . . . The people who had worked with relay circuits were, of course, aware of how to make these things. But they didn't have the mathematical apparatus of the Boolean algebra.

Every single concept from Boole's algebra had its physical counterpart in an electric circuit. An on switch could stand for "true" and an off switch for "false," and the whole thing could be represented in 1's and 0's. More important, as Shannon pointed out, the logical operators of Boole's system—AND, OR, NOT—could be replicated exactly as circuits. A connection in series becomes AND, because the current must flow through two switches successively and fails to reach its destination unless both allow it passage. A connection in parallel becomes OR, because the current can flow through either switch, or both. The current flows through two closed switches in parallel and lights a light; $1 + 1 = 1$.

A leap from logic to symbols to circuits: "I think I had more fun doing that than anything else in my life," Shannon remembered fondly. An odd and wonkish sense of fun, maybe—but here was a young man, just twenty-one now, full of the thrill of knowing that he had looked into the box of switches and relays and seen something no one else had. All that remained were the details. In the years to come, it would be as if he forgot that publication was something still required of brilliant scientists; he'd pointlessly incubate remarkable work for years, and he'd end up in a house with an attic stuffed with notes, half-finished articles, and "good questions" on ruled paper. But now, ambitious and unproven, he had work pouring out of him.

Finished in the fall of 1937, Shannon's master's thesis, "A Symbolic Analysis of Relay and Switching Circuits," was presented to an audience in Washington, D.C., and published to career-making applause the following year. In his new system, Shannon wrote in his driest scientific prose,

> any circuit is represented by a set of equations, the terms of the equations corresponding to the various relays and switches of the circuit. A calculus is developed for manipulating these equations by simple mathematical processes, most of which are similar to ordinary algebraic algorithms. This calculus is shown to be exactly analogous to the calculus of propositions used in the symbolic study of logic. . . . The circuit may then be immediately drawn from the equations.

That was the key—after Shannon, designing circuits was no longer an exercise in intuition. It was a science of equations and shortcut rules. Consider a problem that Shannon's colleagues might have faced as they worked to subject their huge analog machine to electric controls. Say that a certain function in the circuits would allow the current to pass through—would output a "1," in Shannon's terms—depending on the state of three different switches, *x*, *y*, and *z*. The current would pass through if only *z* were switched on, or if *y* and *z* were switched on, or if *x* and *z* were switched on, or if *x* and *y* were switched on, or if all three were switched on. Through trial and error, Shannon's colleagues could sooner or later rig up eleven separate connections that would do the job. But Shannon would start with pencil and paper, and his ubiquitous writing pad. He would write out the function in Boolean terms, like this:

$$x'y'z + x'yz + xy'z + xyz' + xyz$$

Then he would boil it down. Two terms in that function feature *yz*, and two feature *y'z*, so he would just factor them out as in any algebra problem:

$$yz(x+x') + y'z(x+x') + xyz'$$

But Boole's logic tells us that $x + x'$ is always true, which makes sense: *x* is either true, or it isn't. Shannon could recognize, then, that $x + x'$ wouldn't really tell him anything interesting about the output of the circuit, and so it could be safely crossed out:

$$yz + y'z + xyz'$$

Now two terms feature *z*, so Shannon could factor again:

$$z(y + y') + xyz'$$

And for the same reason as before, he could cross out the terms in parentheses:

$$z + xyz'$$

There was one more rule in Boole's logic that could distill this even further. Boole showed that $x + x'y = x + y$, or to put it into plain English, asking for a Londoner who was either blue-eyed, or left-handed but not blue-eyed, was just the same as asking for a Londoner who was either blue-eyed or left-handed. Using that rule on the function above, Shannon could cross out z' as redundant, leaving him with this:

$$z + xy$$

Remember the muck in which Shannon had started. His math was able to prove that these two sets of instructions are *exactly the same:*

> Turn on if only *z* is switched on, or if *y* and *z* are switched on, or if *x* and *z* are switched on, or if *x* and *y* are switched on, or if all three are switched on.
>
> Turn on if *z* is switched on, or if *x* and *y* are switched on.

In other words, he had discovered a way to do the work of eleven connections with just two, a parallel and a series. And he had done it without ever touching a switch.

Armed with these insights, Shannon spent the rest of his thesis demonstrating their possibilities. A calculator for adding binary numbers; a five-button combination lock with electronic alarm—as soon as the equations were worked out, they were as good as built. Circuit design was, for the first time, a science. And turning art into science would be the hallmark of Shannon's career.

One more beauty of this system: as soon as switches are reduced to symbols, the switches no longer matter. The system could work in *any* medium, from clunky switches to microscopic arrays of molecules. The only thing needed was a "logic" gate capable of expressing "yes" and "no"—and the gate could be anything. The rules for easing the labor of a room-sized mechanical computer are the same rules that would be built into the circuits of vacuum tubes, transistors, microchips—at every step, the binary logic of 0 and 1.

It was trivial, Shannon said. But it was the kind of discovery blessed to be trivial only after the fact.

Still—"possibly the most important, and also the most famous, master's thesis of the century"? "One of the greatest master's theses ever"? "The most important master's thesis of all time"? "Monumental"? Was a series of time-saving tricks for engineers really worth all of that praise? As long as the job was done either way, was it really so crucial that Shannon could do in two steps what his colleagues did in eleven?

It *was* crucial—but the most radical result of Shannon's thesis was largely implied, not stated, and its import only became clear with time. The implication gets clearer when we realize that Shannon, following Boole, treated the equal sign as a conditional: "if."

1 + 1 = 1: *if* the current passes through two switches in parallel, a light lights (or a relay passes on a signal meaning "yes"). 0 + 0 = 0: *if* the current passes through neither of two switches in parallel, a light fails to light (or a relay passes on a signal meaning "no"). Depending on the input, the same switches could give two different answers. To take an anthropomorphizing leap—a circuit could decide. A circuit could do logic. *Many* circuits could do enormously complex logic: they could solve logical puzzles and deduce conclusions from premises as reliably as any human with a pencil, and faster. And because Boole had shown how to resolve logic into a series of binary, true-false decisions, any system capable of representing binaries has access to the entire logical universe he described. "The laws of thought" had been extended to the inanimate world.

That same year, the British mathematician Alan Turing published a famously critical step toward machine intelligence. He had proven that any solvable mathematical problem could, in principle, be solved by machine. He had pointed the way toward computers that could reprogram their own instructions as they worked, all-purpose machines of a flexibility far beyond any that had yet been conceived. Now, Shannon had shown that any sensical statement of logic could, in principle, be evaluated by machine. Turing's machine was still a construct of theory: he proved his point with a hypothetical "read/write head" operating on an arbitrarily long string of magnetic tape—an imaginary computer with a single moving part. Shannon, on the other hand, had proven the logical possibilities of the circuits that could be found in any telephone switchboard: he had shown, "down to the metal," how engineers and programmers to come might one day wire logic into a machine. This leap, writes Walter Isaacson, "became the basic concept underlying all digital computers."

It would be six more years before Turing and Shannon met in a wartime scientists' cafeteria, each of their projects so classified that they could only speak of them in hints. They had barely begun to build. Yet in one year, "an annus mirabilis of the computer age," they had laid the foundations. In particular, they had shown the possibilities of *digital* computing, of minute, discrete decisions arrayed one after the other. Less than a decade after Shannon's paper, the great analog machine, the differential analyzer, was effectively obsolete, replaced by digital computers that could do its work literally a thousand times faster, answering questions in real time, driven by thousands of logic gates that each acted as "an all-or-none device." The medium now was vacuum tubes, not switches—but the design was a direct descendant of Shannon's discovery.

Yet none of this could have been foreseen in 1937 by Vannevar Bush, planning ever more complex and capable versions of his differential analyzer, or even by Claude Shannon. Beneath the thrum of that remarkable machine, it all might have sounded, in a way, like regress: that the finely engineered discs and gears were to be superseded by switches no more complex in their essentials than a telegraph key; that there was less analytic potential in the 100-ton behemoth than in the small box fastened to its side; that machines so solidly intuitive they could teach an unschooled mechanic calculus by hand were to be replaced by opaque cabinets that gave a blank face to the world. From Thomson to Bush, the analog computer was, in a way, one of engineering's long, blind alleys.

In that light, a story from Hapgood, the MIT historian: "Years ago an engineer told me a fantasy he thought threw some light on the ends of engineering, or at least on those underlying his own labors. A flying saucer arrives on Earth and the crew starts flying over cities and dams and canals and highways and grids of power lines; they follow cars on the roads and monitor the emissions of TV towers. They beam up a computer into their saucer, tear it down, and examine it. 'Wow,' one of them finally exclaims. 'Isn't nature incredible!?' "

Indifferent to beauty as all but survival's side effect, wildly profligate, remorseless: nature and *techne* aren't so different.

5

A Decidedly Unconventional Type of Youngster

It's been said that most of the great writers have bibliographies, not biographies. The kind of life requisite to their work leaves little behind but the words themselves. Even if we had the questionable privilege of watching them scribble for hours every day, we'd find more of who they were simply in the pages of their books. Something similar might be said of Claude Shannon in this period, working with a speed and absorption that he would not match for the rest of his life. What can we recover of who he was from what he made?

Consider some thesis topics chosen by Shannon's contemporaries in MIT's electrical engineering department: "Skin Effect Resistance Ratio of a Circular Loop of Wire"; "An Investigation of Two Methods of Measuring the Acceleration of Rotating Machinery"; "Three Mechanisms of Breakdown of Pyrex Glass"; "A Plan for Remodeling an Industrial Power Plant"; "A Proposal to Electrify a Section of the Boston and Maine Railroad Haverhill Division." All of these were solidly and practically bound to the world of things. In the best tradition of engineering, they found new uses for old materials, or built physical systems to higher standards of efficiency and power.

Next to this good work, Shannon's was different not only in degree but in kind. He was a tinkerer to the end of his life, and he worked with his hands long after he had any need to. But unlike other tinkerers, he had a way of getting behind things. He loved the objects under his hands, right up to the point when he abstracted his way past them. Switches weren't just switches, but a metaphor for math. There had been legions of jugglers and unicycle riders in the world, but few were as compelled as Shannon would be to fit those activities to equations. Most important of all, he would abstract his way past all of human communication, to the structure and form that every message holds in common. In all these endeavors, he was distinguished less by quantitative horsepower than by his mastery of model making: the reduction of big problems to their essential core. In banishing art and ambiguity, in finding the ways in which human artifacts merely stood for mathematics, Shannon's work at twenty-one was a window on all the work he had left.

There are passionate scientists who are almost overcome by the abundance of the world, who are gluttons for facts; and then there are those who stand a step back from the world, their apartness a condition of their work. Shannon was one of this latter kind: an abstracted man. In his twenties—his most productive years—he took his abstraction to the point of deep withdrawal and almost crippling shyness. But an abstracted man might also have it in him to be playful or funny—indeed, might be especially suited for that. To love the things around us and yet to also see them as cheap stand-ins for the real reality of numbers, theorems, and logic might, to the right temperament, give the world the appearance of a permanent joke.

"What's your secret in remaining so carefree?" an interviewer asked Shannon toward the end of his life. Shannon answered, "I do what comes naturally, and usefulness is not my main goal. . . . I keep asking myself, *How would you do this? Is it possible to make a machine do that? Can you prove this theorem?*" For an abstracted man at his most content, the world isn't there to be used, but to be played with, manipulated by hand and mind. Shannon was an atheist, and seems to have come by it naturally, without any crisis of faith; puzzling over the origins of human intelligence with the same interviewer, he said matter-of-factly, "I don't happen to be a religious man and I don't think it would help if I were!" And yet, in his instinct that the world we see merely stands for something else, there is an inkling that his distant Puritan ancestors might have recognized as kin.

Something in Shannon—perhaps just this withdrawn unworldliness—seemed to trigger the protective instincts of others, even from the generally unsentimental technicians of MIT. Rail-thin, small-town, transparently brilliant; a face made of angles, and an Adam's apple too large for his neck: he must have looked like the kind of young man always on the verge of being mugged or hit by a bus. When he enrolled in a flying class in the wake of his thesis's publication, the MIT professor teaching the course immediately marked him out as odd—odd even for Cambridge—and canvassed his colleagues for their opinions. From his extracurricular investigations, the flight instructor wrote in a letter to MIT's president, "I am convinced that Shannon is not only unusual but is in fact a near-genious [*sic*] of most unusual promise." With the president's permission, he would ban Shannon from the cockpit: such a life wasn't worth risking in a crash.

Two days later, the president, physicist Karl Taylor Compton, sent back a levelheaded reply: "Somehow I doubt the advisability of urging a young man to refrain from flying or arbitrarily to take the opportunity away from him, on the ground of his being intellectually superior. I doubt whether it would be good for the development of his own character and personality."

So with the endorsement of the administration, Shannon kept flying: like any other student, he was permitted to risk the contents of his brain. He risked his in the flight school's simple propeller crafts, blades buzzing like an overgrown wasp, and he always came down safely. A 1939 photo shows him standing beside a Piper Cub, a light two-seater popular with flight schools. He's incongruously well dressed, his white collar well starched and his tie tightly knotted, and he addresses the camera seriously as he rests his hand on the plane's propeller.

Those responsible for Shannon's career were nearly as protective as those responsible for his safety. Bush described him to a colleague as "a decidedly unconventional type of youngster. . . . He is a very shy and retiring sort of individual, exceedingly modest, and who would readily be thrown off the track." But even had it been clear that Shannon's thesis prophesied the end of the analog computing to which his advisor had devoted a decade and a half, Bush was a teacher and engineer large-spirited enough to recognize brilliance when he saw it. As science writer William Poundstone notes, "Bush believed Shannon to be an almost universal genius, whose talents might be channeled in any direction." More than that, Bush took it upon himself to choose the direction.

Bush was, by the late 1930s, one of the most powerful figures in American science, and Shannon was fortunate to have won him for an advocate. The year Shannon's thesis was published, Bush impressed on him that mathematics, not electrical engineering, was the higher-prestige field, and he sponsored Shannon's acceptance into MIT's doctoral program for mathematics. At the same time, Bush's influence in the engineering world won Shannon's thesis the unfortunately named Alfred Noble Prize (unfortunately named, because this is the point at which every writer mentioning it points out that it has no relation to Alfred Nobel's much more famous prize). Awarded by America's engineering societies for the best paper by a scholar under thirty, the Noble meant early distinction within the field, an engraved certificate, and a $500 stipend. It also meant some modest recognition outside the field, including a brief notice—"Youthful Instructor Wins Noble Award"—on page 8 of the *New York Times*. Back in Michigan, the *Otsego County Herald Times* hailed Shannon as a local boy made good (on the front page, naturally).

When news of the award reached Shannon, he knew whom to thank. "I have a sneaking suspicion that you have not only heard about it but had something to do with my getting it," Shannon wrote to Bush. "If so, thanks a lot."

Finally, Bush took it upon himself to find a suitable dissertation project for Shannon in the field of—genetics. Genetics? It was at least as plausible an object for Shannon's talents as switches. Circuits could be taught, genes could be taught—but the analytic skill it took to find the logic beneath them seemed more likely to be inborn. Shannon had already used his "queer algebra" to great effect on relays; "another special algebra," Bush explained to a colleague, "might conceivably handle some of the aspects of Mendelian heredity." More to the point, it was a matter of deep conviction for Bush that specialization was the death of genius. "In these days, when there is a tendency to specialize so closely, it is well for us to be reminded that the possibilities of being at once broad and deep did not pass with Leonardo da Vinci or even Benjamin Franklin," Bush said in a speech at MIT. "Men of our profession—we teachers—are bound to be impressed with the tendency of youths of strikingly capable minds to become interested in one small corner of science and uninterested in the rest of the world. . . . It is unfortunate when a brilliant and creative mind insists upon living in a modern monastic cell."

The words predate Shannon's arrival in Cambridge, but they could have easily expressed Bush's ambitions for his student. And so Shannon was to leave the monastic cell of the differential analyzer (filled, like a monastery, with shifts of men keeping quiet watch at all hours) and the even smaller cell of the circuit box, to go 200 miles south to Cold Spring Harbor on Long Island, and to come back with a dissertation. If any protest came from Shannon, it was not recorded.

6

Cold Spring Harbor

In the summer of 1939, then, Shannon arrived at one of the greatest genetics laboratories in America, and one of the greatest scientific embarrassments: the Eugenics Record Office. In 1910, when this headquarters of the American eugenics movement opened its doors, it was considered the cutting edge of progress in some circles to push for the selective breeding of the "fittest families" and the sterilization of the "defective classes." Its founder had opined that "three or four per cent of our population is a fearful drag on our civilization," and its longtime director even mailed state legislators custom-made estimates of the number of "defectives" in their districts. By 1939, the movement was dying, and the crimes of Nazi Germany—a regime that took eugenicists as seriously as they took themselves—were the final push into disrepute. (Chillingly enough, a Nazi poster from 1936 prominently featured the flag of the United States, along with those of other nations that had adopted eugenics laws. The inscription read: "WE DO NOT STAND ALONE.") Somewhere on the list of Vannevar Bush's accomplishments, then, should be his role in killing American eugenics. As president of the Carnegie Institution of Washington, which funded the Eugenics Record Office, he forced its sterilization-promoting director into retirement and ordered the office to close for good on December 31, 1939.

But the poison tree bore some useful fruit. And Shannon was there, in its last months, to collect what he could of it. Few scientists had compiled better data on heredity and inheritance than eugenicists—in some ways, eugenics are to modern genetics as alchemy is to chemistry, the disreputable relative in the attic. Perhaps the best data set of all belonged to the Eugenics Record Office, which had spent a quarter century accumulating more than one million index cards bearing information on human traits and family trees.

Many of those cards were the work of generations of field researchers; many more were volunteered by the subjects themselves, offered freely in return for advice on the fitness of their offspring. In a massive, fireproof vault the cards stood in their files row after row: traits of physiology ("biochemical deficiencies, color blindness, diabetes"), of personality ("lack of foresight, rebelliousness, trustworthiness, irritability, missile throwing, popularity, radicalness, conservativeness, nomadism"), of social behavior ("criminality, prostitution, inherited scholarship, alcoholism, patriotism, 'traitorousness' "), and on and on. Each trait was coded like a book in a library. Searching for chess-playing ability would take Shannon to file 4598: 4 for mental trait, 5 for general mental ability, 9 for game-playing ability, and 8 for chess.

Mixed almost at random through this genetic Library of Babel was good, hard data; junk (unreliable testimony of untrained volunteers; exhaustive reports on circus freaks; "Midget Schedules"); and the bastard mush in between. For an example of the latter, erring on the side of junk, consider the observations of the office's founder on "thalassophilia," or the genetic love of the sea that allegedly causes sailing careers to be passed down through family trees: "Sometimes a father who shows no liking for the sea . . . may carry a determiner for sea-lust recessive. It is theoretically probable that some mothers are heterozygous for love of the sea, so that when married to a thalassophilic man half of their children will show sea-lust and half will not."

While Vannevar Bush might have appreciated this line of thought, descended as he was from several generations of thalassophilic sea captains, much of the junk here lies in the simple-minded assumption that such a complex trait, if it had any basis in genetics at all, could be controlled by a single gene. But this was the murk in which the study of genetics found itself without serious math, and with still more than a decade to go until the spiral staircase of DNA first appeared under biologists' X-rays. Pending the proof, Shannon wrote, we can only speak "as though the genes actually exist." More to the point, without the application of statistics and probability to huge numbers of traits across entire populations, genetics would fail to account for anything more interesting than the height of pea plants or the shape of rooster combs. Eugenicists would be stuck in fruitless, and dangerous, speculations about the gene for sea-lust or for traitorousness. In Shannon's childhood, scientists like J. B. S. Haldane, Ronald Fisher, and Sewall Wright had begun to train the big guns of statistics on biology, effecting the "great modern synthesis" between Darwinian evolution and Mendelian genetics, of which Darwin had been ignorant. It was their work that gave an unexpected value to the raw data vaulted in a selective-human-breeding warehouse, and this was why Claude Shannon was pulled out of the differential analyzer room and enlisted to continue their work in population genetics. Demand for naturalists and butterfly nets had cratered; biology, like computer building, demanded mathematicians.

Long before she helped to remake the study of genetics, Barbara Stoddard Burks, Shannon's supervisor at Cold Spring Harbor, had narrated a children's picture book: "Thousands of stars glittered in the heavens, and Father showed me the Southern Cross, which is made of four very bright stars in the shape of a kite. Grown people call the shape a cross, though, and some feel very proud when they have seen it because they have to travel so far before they can."

Few scientists had traveled as far as Burks. As a child, she'd traveled with her parents, two educators, to the Philippines' remote mountains, and when she returned to America, she starred in a picture book, *Barbara's Philippine Journey,* written by her mother in young Barbara's voice. She had traveled to the upper ranks of American science, even at a time when women were still shunted out of the allegedly purer disciplines, and from theory to fieldwork. Like Shannon, fourteen years her junior, she did her best work in her twenties; unlike him, she was a woman who had learned to cope with colleagues who stigmatized her as exceedingly aggressive for defending her conclusions with the same confidence they owned.

Burks had traveled with her field, bringing statistical rigor to the study of genetics. Much of her work was directed at the age-old nature-nurture problem, particularly with respect to intelligence. Burks's most controversial studies were efforts to isolate the impact of genetics and environment on IQ. Nature without nurture, for instance, was a study of identical twins raised apart; nurture without nature was a comparison of the intelligence of foster children and their foster parents. At twenty-four, her study of foster children led her to the contentious conclusion that variance in IQ was 75 to 80 percent inherited. Though Burks had no truck with eugenics, the million index cards at Cold Spring Harbor brought her there for the same reason they attracted Bush's attention, and in the eugenics office's last years, she published a reliable method of filtering out the garbage in the files to get at the usable data.

Burks was, in other words, both an expert on and a model of intelligence; so her words carried some weight when, after reading some of his preliminary work on genetics, she wrote back to MIT that "surely Shannon is gifted—perhaps to a *very* high degree." And she shared a moment of commiseration with Bush over this young man who seemed to have so little to learn from either of them: "To advise a youth like Shannon is difficult, is it not?" All the same, Shannon still had to learn the entire field of genetics from scratch. Alleles, chromosomes, heterozygosity—when he first sat down to it, he confessed to Bush, he didn't even understand the words. From this impoverished start, he (mostly) mastered a new science and produced publishable work in less than a year.

"An Algebra for Theoretical Genetics" did, in fact, bear all the marks of a gifted novice airdropped into strange territory—for better and worse. Shannon only bothered to cite seven other studies in his bibliography, excusing himself on the grounds that his method of genetic math was literally unprecedented: "No work has been done previously along the specific algebraic lines indicated in this thesis." But this faith in his own originality cost him: at one point, he presented as a new discovery a theorem that had been common knowledge among biologists for two decades. One class in genetics, or a few more weeks in the library, might have saved him the trouble of rediscovering it from scratch. As he confided to Bush after the fact, "Although I looked through the textbooks on Genetics fairly carefully, I didn't have the courage to tackle the periodical literature." At the same time, though, Shannon offered genuinely new eyes on old problems, and where his thought *was* original, it was almost unconsciously so. Like something of a genetic Joseph Conrad, he could reach heights of creativity in an adopted language because he had missed learning its clichés in his youth.

Shannon's genetic algebra was, in effect, an attempt to re-create for cells what he had accomplished for circuits. Circuits before Shannon might be drawn on a blackboard, but not represented as equations. Of course, it's much more unwieldy to manipulate a diagram than an equation, and one couldn't even begin to use mathematical rules on a drawing. Everything in Shannon's thesis flowed from his realization that circuits were *poorly symbolized*. What if genes were poorly symbolized, too? Just as Boolean algebra helped automate the mental effort of wiring machines, an algebra for genetics might help biologists predict the course of evolution. The trick, as before, was abstracting away from what was in front of his eyes. Forget the hundred switches in the box; forget that 4598 means chess playing.

"Much of the power and elegance of any mathematical theory," Shannon wrote, "depends on use of a suitably compact and suggestive notation, which nevertheless completely describes the concepts involved." It was a point, in fact, that was already well hammered into the heads of mathematicians, who learned early on, for instance, how Newton and Leibniz had invented the calculus almost simultaneously, but how Leibniz's system of symbols had won out as the more intuitive. But what would be an intuitive system of symbolizing an entire population down to the genes?

As Shannon had learned in the months before sitting down to write, "genes are carried in rodlike bodies called *chromosomes,* a large number of genes lying side by side along the length of a chromosome." (The chromosomes are themselves made of DNA molecules that encode genes in a four-"letter" alphabet, though no one knew that yet.) In most species more complex than a single cell, individuals have a number of pairs of chromosomes (we humans have twenty-three pairs). In species that reproduce sexually, one comes from the mother and one from the father. To simplify, Shannon suggested that we consider an organism with just two chromosome pairs and sixteen genes. He symbolized its genetic code like this:

$$\mathbf{A_1}\,B_1\,C_3\,D_5 \quad E_4\,F_1\,G_6\,H_1$$
$$\mathbf{A_3}\,B_1\,C_4\,D_3 \quad E_4\,F_2\,G_2\,H_2$$

The top left entry, $A_1\ B_1\ C_3\ D_5$, is the chromosome from one parent, and the bottom left entry, $A_3\ B_1\ C_4\ D_3$, is the chromosome from the other; together they make up one chromosome pair. The column of A_1 and A_3 (which is bolded) makes up a gene position. Taken individually, A_1 is an allele, or a gene from one parent for a certain trait. A limited number of alleles is possible at any gene position, and the interaction of alleles from mother and father determines the qualities their offspring inherit. Shannon symbolized the possible alleles with the numbers in subscript. A_1 and A_3 are different expressions of the same trait (hair color, for instance—one for brown and one for blond), and the quality that prevails depends on which gene dominates the other.

Now simplify even further: Imagine that we want to study an entire population of individuals with respect to just two of its traits, A and B. Again, each row of symbols comes from one parent, and each column represents a gene position. Say that there are two possible alleles for A (brown hair and blond hair, for instance), and three for B (tall, medium, and short, for instance). In that case, there are twenty-one genetically distinct individuals (trust us on this), ranging from

$$A_1\,B_1$$
$$A_1\,B_1$$

to

$$A_1\,B_3$$
$$A_2\,B_2.$$

So how could we simulate that population's genetic change over time, or predict the results if it bred at random with another group? What would the new population look like in five generations? In a thousand generations?

If we had an infinity of paper and patience, we might do the math separately for each of the twenty-one individuals, combined randomly with individuals from the crossbreeding group. That would give us one generation, and we could repeat the process over and over until we gave up. But what if the entire population and all of its relevant genes could be represented by just one algebraic expression? The expression would have to be, as Shannon said, both compact and suggestive: compact enough to use as a single variable in an equation, and suggestive enough to be "unpacked" to all of its constituent individuals whenever we wanted to halt the cycles of recombination and investigate the results.

Reasoning like that, Shannon invented a symbol to sum up an entire population: λ^{hi}_{jk}.

That expression is really, as he observed, "a whole group of numbers." λ is the population as a whole. h, j, i, and k are genes. As we come to know the range of genes possible for the population, we can replace those letters with a range of numbers. The column ${}^{h}_{j}$ is one gene position, and because the first trait under consideration has two alleles, the value of h or j can range from 1 to 2. The column ${}^{i}_{k}$ is the other gene position, and because the second trait under consideration has three alleles, the value of i or k can range from 1 to 3. λ^{13}_{22} now stands not for a single individual, but for a fraction of the entire population bearing the genetic code

$$\begin{matrix} A_1 B_3 \\ A_2 B_2. \end{matrix}$$

λ^{hi}_{jk} is an especially elegant way of symbolizing gene frequencies, because, like any good optical illusion, it reveals two different sets of information depending on the way we read it. Read vertically, the columns of variables—${}^{h}_{j}$ and ${}^{i}_{k}$—represent gene positions, which lead us to the qualities of any individual in the population. Read horizontally, the rows of variables—*hi* and *jk*—represent sets of chromosomes, each the inheritance of one parent.

This was, in other words, Shannon's attempt to re-create the central conceptual leap of this thesis on circuits. As before, a wise choice of symbols—addition for a parallel circuit, or a grid of variables for chromosomes—would allow Shannon to simplify and simulate the future on paper. The rest of his dissertation was a set of genetic theorems that put his algebraic tools to work. He could estimate the probability that a gene would appear in an individual after *n* generations of mating. He could use addition to stand for the combination of several populations, and multiplication to stand for random breeding, and he showed how to calculate the product of two populations, $\lambda^{hi}_{jk} \bullet \mu^{hi}_{jk}$. There were fractions of populations, imaginary "negative populations," and rates of change in gene frequencies over time. He could consider "lethal factors," or natural selection against maladaptive traits over time: evolutionary algebra. There were algebraic equations in which *x* was an entire group of organisms: given the genes of a known group in the present, he could work backward and identify the genes of the unknown ancestors who planted its family tree. Most important, he derived the equation—a twelve-line monster of interlocking brackets and exponents—that gave the frequencies of three different alleles in any population after any number of generations. While a number of the dissertation's conclusions were old news, this last result, an extrapolation of the future of any three traits, was entirely new. Less than a year after learning the vocabulary, he had produced findings some five to ten years in advance of the field.

Unlike his discoveries in switching, though, Shannon's genetic work was pitched at far too high a level of abstraction to prove useful. There is some irony in the fact that a facility built for such a practical purpose—to promote the selective breeding of humans—ended with such impractical work. In the case of all but the simplest of organisms, Shannon's algebra demanded too much information to make real-world predictions. "My theory has to do with what happens when you have all the genetic facts," he later explained. "But people don't know all of them, especially for humans. They are pretty well versed on the fruitfly!" Two years after Shannon's death, geneticists finished sequencing the human genome; yet even then, far more input on genetic variation between individual humans would have been required to render Shannon's algebra workable. If anything was to come of Shannon's dissertation, it would not be anything so immediately valuable as a digital computer, but rather new methods and new symbols for thinking through the problems of population genetics in the most general terms.

Even that, though, would have to materialize without Shannon's help. He abandoned his work in genetics as soon as it was typed up and bound.

In a sense, the subject of the dissertation was Claude Shannon himself. The project had been Bush's initiative, and the hypothesis was his. Hypothesis: the subject, a twenty-three-year-old genius, working in a scientific field in which he has no training, in which "he did not even know what the words meant," can produce original findings in less than one year. Conclusion: confirmed, mostly.

Behind the scenes, Bush confidentially canvassed his colleagues for their opinions, admitting as he did that Shannon's work was still marked by amateurism: "It goes on for a while and then just stops, and there are some obvious crudities." He was prepared, then, to put the verdict to Shannon as delicately as possible. "I need your guidance before I speak to him concerning this particular thing," he wrote to a Harvard statistician, "for what I say will either encourage or discourage him greatly." That worry speaks to the touchy pride that Bush saw in his student, "a man who should be handled with great care"—as well as to the simple fact that Shannon's academic life to date, from Gaylord to Cambridge, had been free of failure.

In any case, Bush was spared the work of framing bad news: the reviews came back bearing phrases like "very suitable" and "very much impressed." Burks was even more supportive. The story went that the seventeenth-century mathematician Pascal, at the age of twelve, had independently discovered the theorems of Euclid's geometry by drawing on his playroom floor—and Shannon's work, Burks said, was something like that. "This, I feel strongly, should be polished a bit and then published," Bush wrote to Shannon with some satisfaction.

For all that, Shannon ignored him: his genetics work was filed away and forgotten. There is no sign that Shannon heard any condescension in the comparison to a precocious twelve-year-old scribbling on the floor. At the same time, he seems to have had no wish to be Pascal, rediscovering common knowledge. Such discoveries might say something remarkable about their authors, an unschooled boy or an out-of-place engineer, but they said nothing new about the world. The newest element of Shannon's dissertation was his algebraic method itself—and this would only prove worthwhile if Shannon, a young man with no clout and no network in the field, could convince geneticists to set aside their familiar tools and start using his. Shannon understood that as well as anyone: "I had a good time acting as a geneticist for a couple years," he later joked.

Burks and Bush included, along with their praise for his work, candid assessments of its odds of making an impact. Burks wrote to MIT that "few scientists are ever able to apply creatively a new and unconventional method furnished by someone else—at least of their own generation." Bush passed the warning on to his student, along with the praise: "I doubt very much whether your publication will result in further work by others using your method, for there are very few individuals in this general field who would be likely to do so." The very peculiarity of Shannon's method, the isolation in which he had invented it, would most likely consign it to irrelevance. Or, at the absolute best, it would sentence its inventor to a frustrating career as an outsider geneticist peddling his notation to skeptics. For a student who had already made his name as one of the nation's most talented young engineers, it must have looked like an unappealing future—and an unnecessary one. Shannon, observed a later colleague, "did not need to corrupt his reputation with anything non-spectacular."

Shannon would be adamant on the point, for any number of publications-that-might-have-been, for the rest of his life: after the effort of discovery, the effort of communication was secondary, by far. He had solved a problem to his own satisfaction—and that, as far as he was concerned, was enough, especially in the sub-spectacular cases. Shannon explained later: "After I had found the answers it was always painful to write them up or to publish them (which is how you get the acclaim)." A more magniloquent scientist might have added something about the pure Platonic joy of discovery. Not Shannon, though: "Too lazy, I guess."

More than a half century after the dissertation was submitted to Bush and Burks, the editors of Shannon's collected papers asked an expert in modern population genetics to read Shannon's lost dissertation with a counterfactual eye: if it had been published, and if it had been read, would it have mattered? The reviewer compared the dissertation to the work of two other young, mathematically inclined geneticists who were also working in obscurity in the late 1930s. While he ranked Shannon as the least of the three, he admitted his regret "that the work of all three did not become widely known in 1940. It would have changed the history of the subject substantially, I think."

Shannon would have to make his history elsewhere. The simplest explanation for his failure to publish is just that his attention did what it did so often: it wandered away. In the midst of what was supposed to be his immersion in genetics, he stopped to write a letter to his advisor:

> *Dear Dr. Bush . . .*
>
> *I've been working on three different ideas simultaneously, and strangely enough it seems a more productive method than sticking to one problem. . . .*
>
> *Off and on I have been working on an analysis of some of the fundamental properties of general systems for the transmission of intelligence, including telephony, radio, television, telegraphy, etc. . . .*

7

The Labs

Real life mathematics . . . requires barbarians: people willing to fight, to conquer, to build, to understand, with no predetermined idea about which tool should be used.

—Bernard Beauzamy

That would have to wait. Not even the most important master's thesis ever written and publishable work in genetics were enough for a PhD. Like every other MIT student, Claude Shannon had to pass his mandatory language exams. So he returned to Cambridge, and in between his teaching duties in the mathematics department and his first sketches on telegraphs, telephones, radio, and TV—whatever a mathematician might have to say about four means of communication that had precious few "fundamental properties" in common—he wrote out his stacks of flash cards. French was easier; he failed German before passing on the second try.

In the midst of a thoroughly numeric life, his recreations were the opposite. He developed a passion for jazz, especially for the improvisations he called "unpredictable, irrational." In Gaylord, he played brass horn in the marching band; in Cambridge, jazz clarinet in his room. Playing backup to his record collection was his main distraction from the "intelligence" project, which increasingly cost him late nights and late mornings rolling out of bed. He put up with two roommates in an apartment at 19 Garden Street, not far from Harvard Square. We can imagine him forced up from his desk by the low roar of conversation whenever they threw a party, unpracticed in small talk, fond of walls and doorways. In fact, he was standing in his doorway at just such a party when a popcorn kernel hit him in the face.

Norma Levor, who threw the popcorn to get the attention of the tall, silent young man in the doorway, was only nineteen, but she was easily the most worldly person Shannon had ever known. She was born in a penthouse on New York's Central Park West, her mother the heir to a pincushion fortune, her father an importer of fine Swiss fabrics. Tutelage in Upper West Side left-wing politics from her cousin, a "red" Hollywood screenwriter and playwright, and her sister at Columbia Law (where the radical students were Trotskyists, she said, and the mainstream students were just regular communists), then a summer in Paris as a reporter before her parents brought her home in advance of war ("That's why I was there," she told them, but they wouldn't listen), then on to study government at Radcliffe, and then on to this terribly boring party, where a gaunt young man was standing at the edge of his bedroom, listening to his own jazz on his own record player.

"Why don't you come out here where everybody is?" she asked. He answered, "I like it here, got some great music."

"Bix Beiderbecke, you got him?"

"My favorite."

And that was that. Norma was drawn, she recalled, to Claude's "Christ-like" looks. Christ by way of El Greco, maybe, stretched out across a long frame—but Norma had good taste in most things. And Claude had twenty-four-hour access to the differential analyzer room. It was the scene of much of their courtship, compressed as it was—too short, probably, for Norma to weigh the costs of leaving school, or to discover anything in Shannon's personality other than the odd genius that was his universal first impression, and so short that Claude's entire experience of romance before marriage was the first unstable flush of early-twenties love. Liberal and uninhibited as Norma was, she was the obverse of everything he had left behind in leaving Gaylord. "We spoke to each other in our own private intellectual-silly language," she wrote. "He loved words and repeated 'Boolean' over and over for the sound of it." He wrote her poems, some naughty, all lowercased in the style of e. e. cummings. She said she was a third-generation atheist. He replied, "How can you be anything else?"

They were inseparable to begin with, to the point that Norma ran into "big trouble" sneaking back in the mornings to her Radcliffe dorm. In the beginning, Claude "was so loving and so darling and so funny and so sweet, so full of fun and such a joy to be with, so great all the time, night and day for months and months and months." The popcorn hit him in the face in October 1939; January 10, 1940, was their wedding day, in a Boston courthouse with a justice of the peace. The honeymoon in New Hampshire was only marred by an anti-Semite hotel keeper who denied them a room (Norma was Jewish; Claude apparently looked it).

Shannon seemed pleasantly befuddled with the speed with which it had all happened. He wrote to Bush, "I did not, as you may have anticipated, marry a lady scientist, but rather a writer. She was helping me with my French (?) and it apparently ripened into something more than French."

In the spring, he put on cap and gown to celebrate his simultaneous master's and doctoral degrees, and the National Research Fellowship he had won, with Bush's help, to spend the following academic year at the famous Institute for Advanced Study (IAS) in Princeton. Asked how the prestigious fellowship came about, he was more than usually sarcastic: "Well, I applied for it and that's how it came, you applied for these things. Tell them how great you are, how smart you are." Norma left her senior year at Radcliffe to follow him—not an unusual decision for a wife in those days, but one that would prove progressively more galling. In her own fields of left-wing politics and writing, Norma's intellectual ambitions were a match for her husband's, but they were put on hold.

Before Princeton, though, the pair would have a brief summer's stop in Norma's childhood home: Manhattan. The summer of 1940 was Claude's second invitation to the Bell Laboratories. Now, though, he returned not as a first-year graduate student but as an award-winning PhD with Vannevar Bush as a patron. He was headed to what was perhaps the world's foremost technology company—and the home of the best communications minds in America.

Had he wanted, Shannon could have continued the glide path through academia, collecting fellowships, amassing awards, and working his way to tenure and a lifetime of professorial comfort. But Shannon had proven himself the kind of mathematician who could stand on his own two feet outside the academy, whose work might result in more than a university chair. Shannon's foremost mentor, Bush, understood this, too, and he set out to shape the course of Shannon's life accordingly.

It helped, of course, that Vannevar Bush was, at the time, the high priest of applied mathematics. He may not have groomed Shannon explicitly in his own image, but he understood that Shannon's talents, properly harnessed, would serve him well outside of a university setting, the same way Bush's talents had taken him to a position of national prominence. It was Bush who had hired Shannon to work on the differential analyzer; Bush who pushed Shannon to apply his studies in mathematical logic to theoretical genetics; and Bush who, in 1938, put Shannon to work on the microfiche rapid selector, a "light-sensing reader system to allow speedy retrieval of microfilmed information"—a far cry from any of Shannon's graduate work, but yet another chance to force his student to flex his mathematical muscles in an unfamiliar domain. Bush, too, had a tinkerer's instinct, and he set to work on Shannon as a tinkerer might: a fresh problem here, a new research topic there, and eventually Shannon would be transformed into an applied mathematician of the first rank.

After his acceptance to the Institute for Advanced Study, but before leaving for Bell, Shannon wrote to Bush seeking career advice. Bush was emphatic: "The only point I have in mind is I feel that you are primarily an applied mathematician, and that hence your [research] problem ought to lie in this exceedingly broad field rather than in some field of pure mathematics."

But Bush wasn't the only one who understood that Shannon's true potential lay somewhere other than pure math. Thornton C. Fry, head of the Bell Labs mathematics group, had taken notice as well. Fry was "a very careful and formal person." That was the charitable way of saying he was a stiff: while working at the National Center for Atmospheric Research, he "rather frowned on the informal western clothing of the NCAR staff," though "this never influenced his respect for their work." Fry's manner reflected his roots as the son of an Ohio carpenter. By 1920, he had managed to escape the family trade and had finished a tripartite PhD in mathematics, physics, and astronomy.

It was a combination of luck and skill that helped Fry turn that training into a job at Western Electric, AT&T's equipment manufacturer and one of the country's leading engineering organizations. Interviewed by Western Electric's research chief, Fry was caught unusually flat-footed by the questions. He wanted to know: how familiar was Fry with the work of the era's most influential communications engineers? As Fry later recalled his fiasco of an interview: "Had I ever read the works of Heaviside? I'd never heard of Heaviside. . . . He asked me if I had ever heard of Campbell. I'd never heard of Campbell. I think he asked me if I'd ever heard of Molina. I hadn't. Whatever he asked me, I hadn't." Still, something about this too-formal young man was impressive; Western Electric rolled the dice and gave Fry the job. He excelled at it, and after the Western and AT&T research divisions were spun off to form Bell Labs, Fry found himself running the Labs' mathematics research group.

Bell Labs "was where the future, which is what we now happen to call the present, was conceived and designed," wrote Jon Gertner in *The Idea Factory,* his history of the Labs. Other appraisals struck a similar note: "the crown jewel"; "the country's intellectual utopia." By the time Shannon joined Bell Labs, the curious mix of techniques, talent, culture, and scale had turned the modest R&D wing of the phone company into a powerhouse of discovery. It was an institution that churned out inventions and ideas at an unheard-of rate and of unimaginable variety. In Gertner's words, "to consider what occurred at

Bell Labs . . . is to consider the possibilities of what large human organizations might accomplish."

Its founder was a tinkerer of an earlier era: Alexander Graham Bell. United States Patent No. 174,465—for "the method of, and apparatus for, transmitting vocal or other sounds telegraphically . . . by causing electrical undulations, similar in form to the vibrations of the air accompanying the said vocal or other sound"—earned Bell the title "inventor of the telephone," worldwide recognition, and a considerable fortune. He founded a phone company, American Telephone & Telegraph (AT&T), whose goal was suitably immodest: turn Bell's invention into a nationwide network of phones, lines, and transmitters. The result: within a decade, the telephone went from lab demonstrations to a fixture in 150,000 American homes. By 1915, the network was a marvel of human engineering, a continent-spanning web that allowed for transcontinental communication at a time when physical travel from coast to coast still took nearly a week.

In 1925, Bell Labs was carved out of the phone company as a stand-alone entity, with custody shared jointly by AT&T and Western Electric. Walter Gifford, the president of AT&T, observed that the Labs, while nominally an arm of the phone company, could "carry on scientific research on a scale that is probably not equaled by any organization in the country, or in the world." The goal of Bell Labs wasn't simply clearer and faster phone calls. The Labs were tasked with dreaming up a future in which every form of communication would be a machine-aided endeavor.

So-called basic research became the Labs' lifeblood. If Google's "20 percent time"—the practice that frees one-fifth of a Google employee's schedule to devote to blue-sky projects—seems like a West Coast indulgence, then Bell Labs' research operation, buoyed by a federally approved monopoly and huge profit margins, would appear gluttonous by comparison. Its employees were given extraordinary freedom. Figure out, a Bell researcher might be told, how "fundamental questions of physics or chemistry might someday affect communications." *Might someday*—Bell researchers were encouraged to think decades down the road, to imagine how technology could radically alter the character of everyday life, to wonder how Bell might "connect all of us, and all of our new machines, together." One Bell employee of a later era summarized it like this: "When I first came there was the philosophy: look, what you're doing might not be important for ten years or twenty years, but that's fine, we'll be there then."

The extraordinary freedom was a scientist's dream, and the ability to work as they pleased drew together an astonishing set of minds. Bernard "Barney" Oliver, a Bell Labs researcher who would later head up research for Hewlett-Packard, recalled thinking, "Gee, you know, here I am, I've got the world's knowledge in electrical engineering at my beck and call. All I've got to do is pick up the phone or go see somebody and I can get the answer."

That accumulation of talent paid tremendous dividends. In the span of a few decades, Bell researchers invented the fax machine, touch-tone dialing, and the solar battery cell. They engineered the first-ever long-distance phone call and synchronized the sounds and images in movies. During the war, they improved radar, sonar, and the bazooka, and they created a secure line to allow Franklin Roosevelt to speak to Winston Churchill. And in 1947, Bell researchers John Bardeen, William Shockley, and Walter Brattain created the transistor, the foundation of modern electronics. The trio would earn a Nobel Prize, one of the six Nobels given to Bell scientists during the twentieth century.

It was one thing for an industrial laboratory to hire qualified PhDs and put them to work on various pressing engineering problems. But Nobel Prizes? Pie-in-the-sky projects? Ten or twenty years of leeway? Even accounting for nostalgia, Thornton Fry's judgment hardly seems out of place; looking back on the Labs, he called it "a fairyland company."

Consider Clinton Davisson, Nobel laureate and Bell Labs researcher. Known as Davy, he was "wraith-like and slow-moving . . . an almost spectral presence." A frail, quiet midwesterner who kept largely to himself, Davy was able to write his own ticket at the Labs. As Gertner put it, "he was allowed to carve out a position as a scientist who rejected any kind of management role and instead work as a lone researcher, or sometimes a researcher teamed with one or two other experimentalists, pursuing only projects that aroused his interest." Importantly, "he seemed to display little concern about how (or whether) such research would assist the phone company."

Bell Labs was neither a university nor a charity. And yet Davy was allowed to conduct endless experiments on the company dime, many of which had only the most tenuous tie to the bottom line. It's telling that Davy's Nobel Prize—awarded for proving that electrons moved in a wave pattern, knowledge gleaned by smashing a piece of crystalline nickel with electrons—won the Labs fame, but no incremental fortune. A mind like his—one that could have navigated its way into the academic career of his choosing—was considered useful to Bell executives, even if the precise use was fuzzy.

A rigorous investment in basic research meant that there were, at any given time, several Davys on the Labs' payroll. Of course, freedom to research at will could be a burden, a kind of anxiety, in its own right. The thinkers who thrived at the Labs were those who, confronted with a nearly limitless field of questions, chose the "right" ones: the ones most fertile of breakthroughs in technique or theory, the ones that opened on broad vistas rather than dead ends. This choice of questions has always been a matter of intuition as much as erudition, the irreducible kernel of art in science.

Claude Shannon was one of those who thrived. Among the institutions that had dotted the landscape of Shannon's life, it's hard to imagine a place better suited to his mix of passions and particular working style than the Bell Laboratories of the 1940s. "I had freedom to do anything I wanted from almost the day I started," he reflected. "They never told me what to work on."

Thornton Fry hadn't simply recruited Shannon to the Labs; he also assigned him to the math group, which Fry had crafted himself to ensure that the talent he recruited wasn't wasted. Fry held strong views about the role of mathematicians within industry, and depending on one's perspective, he was either a visionary or a heretic. In a long, thoughtful meditation published in the *Bell System Technical Journal*, Fry began by pointing out the obvious: there was, for all the enlightened teaching in university math departments, a near-total lack of industrial training for mathematicians who aspired to build things rather than simply think about things. "Though the United States holds a position of outstanding leadership in pure mathematics," Fry wrote, "there is no school which provides an adequate mathematical training for the student who wishes to use the subject in the field of industrial applications rather than to cultivate it as an end in itself."

It's taken as a given in our era that a high-level math mind—a "quant"—can find gainful employment. But that wasn't always the case, and especially not in the world of elite mathematics in the early twentieth century. What was valued in the highest levels of mathematics had precious little application outside of it. Solutions to abstract problems won glory, and thus whole careers were devoted to chasing solutions to problems like the Riemann hypothesis, the Poincaré and Collatz conjectures, and Fermat's last theorem. These were the math world's greatest puzzles, and the fact that decades had passed with no solution made them all the more tantalizing. They were taken dead seriously, and whether or not the solutions had any practical aim or application was an afterthought, if it was a thought at all.

Fry, himself a mathematics PhD, understood this better than most. "The typical mathematician," Fry observed,

> is not the sort of man to carry on an industrial project. He is a dreamer, not much interested in things or the dollars they can be sold for. He is a perfectionist, unwilling to compromise; idealizes to the point of impracticality; is so concerned with the broad horizon that he cannot keep his eye on the ball.

All of which left many a graduate student exceptionally well trained in a style of problem solving that had limited use outside of the mathematical fraternity. An industrial lab, then, had about as much use for a mathematician as a fish had for a bicycle—unless. . . .

Fry's hunch was that not all mathematicians wanted to write papers and chase tenure. He also guessed that the right environment could play to their strengths and put them to work on practical things, set them on "everyday problems" and "concrete exploitation." And he was among the few people in a position to make that happen—and to make his case for the "industrial mathematician" as a new breed of thinker-doer.

He baked his philosophy into the heart of the math group. His case was simple: the engineers of Bell Labs were "pathetically ignorant of mathematics," but math, applied correctly, could help them work out complex problems in telephony. At the same time, the math group served as a catch-all for any gifted member of the Labs staff too odd to play well with others. "Mathematicians are queer people. You are and I am. That's a fact," Fry told a mathematically inclined interviewer. "So that anybody who was queer enough that you didn't know what to do with him, you said, 'This fellow is a mathematician. Let's have him transferred over to Fry.' "

The math group under Fry's leadership began as an in-house consulting organization, with mathematicians available as needed to the engineers, physicists, chemists, and others, but free to pick their own internal "clients." They were there to offer advice and assistance; the management and messy realities of industrial projects could be left to others. As Bell Labs' Henry Pollak observed, "our principle was that we'll do anything once, but nothing twice."

This gave the group a broad mandate, flexible even within the famously loose culture of Bell Labs. As one researcher from that era put it, "Our job was to stick our nose into everybody's business." In Fry's words, "there was nothing that we weren't entitled to work on if we wanted to." Or as Shannon himself recalled, "I was in the mathematics research group which was kind of free-wheeling and not so oriented on projects as people trying to do individual research as fast as they could. . . . I enjoyed it more that way, where I was working on my own projects."

In exchange for its independence, the math group acquainted itself with the phone company's ways. The earliest members climbed telephone poles and operated switchboards. They mastered the mathematics of switching and solved thorny network problems. Like the rest of the Labs' employees, they addressed one another exclusively by last names. In time, their hands-on experience combined with their training would enable them to delve deep into the underlying mathematics of communication engineering. The math group would eventually be regarded as a standout within the industry, and Fry's vision would set the standard for the use of mathematical minds within a large private-sector concern.

Shannon was given a summer's exposure to Bell Labs—and though few records of his summer there exist, we do know something of his output. Shannon's work during this period is captured in two technical memoranda, both of which give a sense of how mathematical skills could meet the telephone company's goals.

Shannon's first effort was the "Theorem on Color Coding." In a system as complex as the Bell telephone network, questions about the coloration of wires were a serious business. Shannon was tasked with finding an answer to the following puzzle:

> There are a number of relays, switches, and other devices A, B, . . . , K to be interconnected. The connecting wires are first formed in a cable with the leads associated with A coming out at one point those with B at another, etc., and it is necessary, in order to distinguish the different wires, that all those coming out of the cable at the same point be differently colored. There may be any number of leads joining the same two points. We might have, for example, four wires from A to B, two from B to C, three from C to D and one from A to D. The four from A to B must all be of different colors, and all different from those from B to C and A to D, but the three from C to D can be the same as three of those A to B. Also the one from A to D can be the same as one from B to C. If we assume that not more than m leads start at any one point, the question arises as to the least number of different colors that is sufficient to color any network.

If this question has the flavor of "two trains leave the station at the same time . . . ," it's because problems like this lend themselves to the search for mathematical shortcuts. That's what Shannon was after here: a workaround, something that would allow Bell engineers without advanced degrees in mathematics a quick and easy way to arrive at the minimum number of colors they needed for a network. And Shannon's answer—multiply the number of network lines by 1.5; the greatest integer less than or equal to that value is the number of colors you need—was thorough and thoughtful and well proved. If it wasn't the stuff of mathematical legend, it was still eminently useful. And unlike, say, an algebra of genetics or a meditation on symbolic logic and circuits, it could be put into immediate practice.

This was significant. The paper illustrates the ways in which the formal education of Claude Shannon the adult had mixed with the informal instruction of Claude Shannon the boy, the one whose childhood was spent happily playing with broken radios and makeshift elevators. And it shows that the part of his nature that was hardheaded and practical had remained firmly intact. It isn't a stretch to imagine that solving this particular problem—technical and narrow though it may have seemed—gave Shannon a great deal of joy. It was, after all, an intricate puzzle. And it was reminiscent, as well, of a youth spent playing telegraph engineer, a kind of graduated version of building a barbed-wire network.

Shannon's second effort, "The Use of the Lakatos-Hickman Relay in a Subscriber-Sender Case," was an attempt to simplify and economize the relays Bell used to connect phone calls. It was the kind of work that called into question whether the Bell network's system of relays, as currently constituted, was optimal, and whether there wasn't a better way to make it operate. In other words, it was a kind of tinkering on the very largest scale, on the beating heart of the phone system. It led Shannon to think up two new options for circuits that drew on his master's thesis work—"designed by a combination of common sense and Boolean algebra methods"—and though he was quick to acknowledge that each of his designs had its own flaws, he also defended them as superior to what was on offer.

When he first arrived at the Labs, Shannon had his doubts: Would an industrial laboratory constrain his ability to think big thoughts and dream up new ideas? After this summer's work, those concerns were put to bed. The Labs had given him as broad a scope as he might have hoped for in a professional setting.

"I got quite a kick," Shannon wrote to Vannevar Bush, "when I found out that the Labs are actually using [my] relay algebra in design work and attribute a couple new circuit designs to it." As with a tinkerer who successfully flips the switch on his latest creation, it isn't difficult to imagine Bush reading that sentence, sitting back, and smiling with satisfaction.

8

Princeton

By the end of his Bell Labs summer and his arrival at the Institute for Advanced Study in Princeton that fall, the name "Claude Shannon" was pinging its way through math and engineering circles. Vannevar Bush had, of course, helped it along. But others were noticing the young mathematician as well. Norbert Wiener, by then no longer a genius in training under his father but a highly respected mathematician in his own right, wrote in 1940 that he thought Shannon "a man of extraordinary brilliancy and intelligence. . . . He has already done work of great originality and is with no doubt a coming man."

On September 27, 1940, Oswald Veblen of the Institute for Advanced Study touted Shannon in a note to Thornton Fry. Veblen saw in Shannon a rare talent and shared a Shannon paper with Orrin Frink, a leader in the mathematical field of topology. At MIT, too, Shannon had been identified as a standout. On October 21, H. B. Phillips, the head of MIT's math department, cabled a fellow faculty member: "Mr. Shannon is one of the ablest graduates we have ever had and can do first class research in any field in which he becomes interested." The recipient of that message was Marston Morse, who had a field of mathematics named after him, and who joined Wiener and Von Neumann as three out of just seven winners of the Bôcher Memorial Prize, one of math's highest honors.

Morse. Phillips. Frink. Fry. Veblen. Bush. By this point, Shannon had acquired an imposing roster of supporters and patrons; these were math's kingmakers, and even without the usual conspicuous striving of the ambitious and talented, he had earned their backing. He had left a mark on men who were discerning judges of raw intellectual horsepower, and they found in him one of their own.

Shuttled up and down the coast, from one elite institution to another, from one set of mentors to another, from fellowship to fellowship: there is sometimes a kind of placelessness that settles into these scientific stories, and Shannon's in particular. In this, the travels of the ambitious young scientist resemble nothing so much as the spiraling journey of the ambitious civil servant of an earlier age, as described memorably by Benedict Anderson:

> He sees before him a summit rather than a centre. He travels up its corniches in a series of looping arcs which, he hopes, will become smaller and tighter as he nears the top. . . . On this journey there is no assured resting-place; every pause is provisional. The last thing the functionary wants is to return home; for he *has* no home with any intrinsic value. And this: on his upward-spiraling road he encounters as eager fellow-pilgrims his functionary colleagues, from places and families he has scarcely heard of and surely hopes never to have to see.

Who were Shannon's new traveling companions in Princeton? Where did they come from?

There was John von Neumann, a Jewish-Hungarian prodigy, who by the age of six could crack jokes in ancient Greek or give you the quotient of 93,726,784 divided by 64,733,647 (or any other eight-digit numbers) without pencil and paper. He was the kind of student who once literally brought a tutor to tears of awe, who spent a college lecture on "unsolved problems in mathematics" doodling the solutions in his notebook. We owe to Von Neumann much of game theory (the formal study of strategic decisions, as in the famous Prisoner's

Dilemma), and much of the intellectual architecture of modern computers, and a decent chunk of quantum mechanics. Shannon called him "the smartest person I've ever met"; it was a common opinion. What began as a relationship of starstruck admiration—"I was a graduate student—he was one of the great mathematicians of the world," said Shannon—would evolve in later years into something more like an equal partnership between two pioneers in the field of artificial intelligence.

There was Hermann Weyl, a refugee from the Nazis, both a mathematician and philosopher of physics. As a mathematician, Weyl worked to reconcile the revolution in quantum mechanics with the doctrines of classical physics. As a philosopher, he considered Einstein's work on the relativity of space and time not only a turning point in science, but a new insight into the relationship between human consciousness and the external world. Just two years after Einstein published his theory of general relativity, Weyl wrote the definitive treatment of relativity's philosophical foundations. "It is as if a wall which separated us from Truth has collapsed," he exulted. "It has brought us much nearer to grasping the plan that underlies all physical happening." This was highly rarefied stuff by Shannon's standards, and it may have been with some trepidation that Shannon sat down in Weyl's office to pitch to his new advisor a research program for the next year.

Weyl was generally dismissive of Shannon's genetics work (as was Shannon himself, by this point), but Shannon could converse fluently on modern physics, and he won Weyl over by developing an analogy between the quantum weirdness with which physicists were grappling and the problems in communications mathematics that he was just beginning to puzzle through. What if the mathematical model for a message sent over telephone or telegraph wires had something in common with the models for the motion of elementary particles? What if the content of any message and the path of any particle could be described not as mechanical motions, or as randomized nonsense, but as random-*looking* processes that obeyed laws of probability—what physicists called "stochastic" processes? Think of "the fluctuations in the price of stocks, the 'random walk' of a drunk in a sidewalk"—think, for that matter, of a clarinet solo—happenings that were less than fixed but more than chance: maybe "intelligence" and electrons were alike in that way, taking haphazard walks within probability's bounds. That got Weyl's attention.

It was one of the early hints that a mathematics of messaging might have something more to say than the most efficient design of a telephone network: that it might offer something more fundamental about "the plan" that the greatest physicists believed they had glimpsed. It was still only a guess, maybe a useful analogy and nothing more; but with Weyl's approval, Shannon brought his full-time attention to bear on the questions of "intelligence" that he had first raised in his letter to Bush.

And, of course, there was Einstein. He had seen his books burned by the Nazis and had read his own name on a list of assassination targets; like Weyl, he had escaped Germany early, and had made his home in Princeton since 1933. There are a few Einstein-and-Shannon stories, and though they surely cannot all be true at the same time, we offer them all in the interest of completeness.

Norma recalled: "I poured tea for [Einstein], and he told me I was married to a brilliant, brilliant man." She elaborated in another interview, when she said that Einstein eyed her over his teacup and remarked, "Your husband has the greatest mind I have ever come across." The anecdote has been repeated often, but it almost certainly never happened. For one thing, by 1940, Shannon had done some interesting, important work, but nothing that would have attracted Einstein's attention. Physics, after all, wasn't Shannon's field. Further, unlike others at the IAS, we have no record of Shannon trying to elbow his way into audiences with the world's best-known and most-sought-after scientists. Nothing in Shannon's behavior would have indicated his interest in subjecting Einstein to a newly minted PhD's thoughts on this or that, and so the scene simply doesn't square with what we know of Shannon. (The more conspicuous John Nash, by contrast, insisted on meeting with Einstein even as a young student and spent an hour walking him through his thoughts on "gravity, friction, and radiation," according to biographer Sylvia Nasar. At the end of the session, Einstein said, "You had better study some more physics, young man.")

On the other hand, a story from Claude's friend and fellow juggling professor, Arthur Lewbel, is more plausible—and suggests that Einstein had more practical interests than the quality of Shannon's mind:

> The story is that Claude was in the middle of giving a lecture to mathematicians in Princeton, when the door in the back of the room opens, and in walks Albert Einstein. Einstein stands listening for a few minutes, whispers something in the ear of someone in the back of the room, and leaves. At the end of the lecture, Claude hurries to the back of the room to find the person that Einstein had whispered to, to find out what the great man had to say about his work. The answer: Einstein had asked directions to the men's room.

Lewbel recalls Shannon sharing this story twice—the only difference being that, in the other version, Einstein was after directions to tea and cookies, an ending that Lewbel confessed was more likely.

Beyond that, Shannon only recalled that he would often pass Einstein on the way to work in the morning, "and he usually would walk along in sort of bedroom slippers and had old clothes hanging on and he looked like a transient almost, I'd say, and I'd go along in my car and I'd wave at him and he'd wave back. He didn't know really who I was but he'd wave back. Probably he thought I was some kind of weirdo."

Beyond everything else, Norma's story of life-altering Einsteinian praise seems hard to reconcile with Claude's version of benign neglect. Whatever their conflicting accounts tell us about Einstein, they shed some light on the two very different people Norma and Claude were discovering in one another, to their growing dismay: theatrical and taciturn, expansive and self-contained. Less than a year into their marriage, it seemed that they had little in common beyond a fondness for jazz.

In truth, the Institute for Advanced Study proved unhealthy for Shannon. For some, it was a land of academic lotus-eating, an island where the absence of the ordinary worries of the job—students, deadlines, publication pressure—proved enervating rather than invigorating. The physicist Richard Feynman, who was working on his doctorate at Princeton while Shannon was at the IAS down the street, observed the inertia firsthand: "A kind of guilt or depression worms inside of you, and you begin to worry about not getting any ideas. . . . You're not in contact with the experimental guys. You don't have to think how to answer questions from the students. Nothing!"

There were only a few months of this for Shannon, rather than a lifetime. He never stagnated in the way that Feynman found all too common among the lifers. But the quiet of the place, and his freedom from obligation, played into his lifelong tendency to isolate himself. Most days were spent shut indoors, alternating between the notepad and the clarinet, and back again. He would hardly even move to turn from math to music—only reposition himself in the chair beside his desk, put on a jazz record, and take up his clarinet to accompany it. Teddy Grace, an earthy southern alto, was his favorite singer:

Turn off the moon, that heavenly spotlight above
Turn off the stars, for I'm falling madly in love. . . .

Norma was isolated, too. A plan to finish college at nearby Rutgers fell through. Cut off from her family and friends in a sleepy college town (especially sleepy after New York and Paris and Boston), she had no desire to be a housewife at twenty—and yet here she was. Norma worked to fill the days. She invited the institute professors over for tea as often as she could, and, on the strength of her fluency in French, found a job with the Economic Section of the League of Nations, which like so many of the academics at her teas had been driven out of Europe by war.

But it wasn't enough. No amount of prodding could make Claude share her passion for politics: "You know where my interests are, and that's enough." Norma grew convinced that Claude was depressed—but, whatever the cause, the marriage ended as quickly as it began.

After the final fight, Norma left Princeton on the train back to Manhattan. Once the divorce was official, she went west, and on the start of her real life: to California, the screenwriting career she'd wanted since she was a child, Communist Party meetings, marriage to a Hollywood fellow traveler, the blacklist, and self-exile to Europe.

Why we love the people we do is one of the enduring human mysteries—surpassed only, perhaps, by the mysterious nature of the stories we tell ourselves about those loves. What, if anything, Claude might have told himself about the end of his relationship with Norma is lost to us. We have the facts: that they married quickly, discovered cracks in their bond in late 1940, and divorced thereafter.

But we know something else, too: we know of the other great struggle in Shannon's personal life during this fraught time. On September 16, 1940, only days after Shannon arrived at Princeton, Franklin D. Roosevelt signed the Selective Service and Training Act, requiring all male citizens from twenty-one to thirty-five to register for the military draft. On October 16, 1940, the mass registration began. The United States had not yet formally joined the war, but the president and his advisors had seen enough of the totalitarian threat to understand its seriousness. On signing the act, Franklin Roosevelt issued a warning: "America stands at the crossroads of its destiny. Time and distance have been shortened. A few weeks have seen great nations fall. We cannot remain indifferent to the philosophy of force now rampant in the world. We must and will marshal our great potential strength to fend off war from our shores."

What those high-flown words meant for the twenty-four-year-old Shannon and men of his generation was the very real possibility that they would be sent overseas to fight a war—a prospect that, up to this point, had still seemed remote. But signing his name to a registration card surely focused Shannon's mind on the grim reality of having to put his research on hold—indeed, put his entire life on hold—and don a uniform.

For Shannon, this was not a welcome prospect. While we have no indication that he went out of his way to avoid the draft, we do know that he was not the least bit eager to deploy overseas. In his own words,

> Things were moving fast there, and I could smell the war coming along. And it seemed to me I would be safer working full-time for the war effort, safer against the draft, which I didn't exactly fancy. I was a frail man, as I am now. . . . I was trying to play the game, to the best of my ability. But not only that, I thought I'd probably contribute a hell of a lot more.

In another interview, he recalled that "if you can make yourself more useful somewhere else you won't get into the Army. That seemed to me a wise move." A friend noted that Shannon, an introvert, worried not only about the dangers of an overseas deployment but also about the close quarters of Army life: "I think he did the work with that fear in him, that he might have to go into the Army, which means being with lots of people around which he couldn't stand. He was phobic about crowds and people he didn't know."

So Shannon turned to his Bell Labs mentor, Thornton Fry, who managed to secure him a contract doing mathematical analysis for the National Defense Research Committee. The leadership of the NDRC was a who's who of the nation's scientists and engineers—and it included most of the key figures in Shannon's professional circle, including the man who had plucked Shannon out of the Midwest: Vannevar Bush.

Bush was the NDRC's godfather. He had firsthand experience of the breakdown in communication between military officers and civilian scientists during World War I, so when he outlined the need for a federal committee to bridge the gap, he spoke with force and conviction. And he carried that conviction into the Oval Office on June 12, 1940, to make the case for the NDRC to the president himself. It took all of ten minutes for FDR to say yes. "There were those who protested that the action of setting up NDRC was an end run," Bush later wrote, "a grab by which a small company of scientists and engineers acting outside established channels, got hold of the authority and money for the program of developing new weapons. That, in fact, is exactly what it was."

For Shannon, the NDRC would represent an end run of a different kind: it freed him from his worries about the draft board. He would, like many mathematicians of his generation, put his mind, rather than his body, to work on the country's behalf.

9

Fire Control

Wartime would interrupt the working lives of a whole generation—but in the universe of possible interruptions, a grant to research national defense issues with some of the nation's foremost engineers and mathematicians was a blessing. Shannon seemed to understand this; it might explain why he tried, in early December, to return the money given to him for his fellowship at the Institute for Advanced Study. But his $166.67 check came back. The "requirements of military training or other defense emergencies" were an exceptional case, the fellowship office observed; Shannon could keep the money, assuming he resumed his research at the war's close.

Thornton Fry had contacted an NDRC colleague, Warren Weaver, to help find Shannon a project. Born in rural Wisconsin, in 1894, Weaver trained at the University of Wisconsin, joined the Army's Air Service in 1917, and, after a detour at Throop College of Technology (later renamed the California Institute of Technology, or Caltech), returned to Wisconsin to teach in and then chair the math department—a department of which Thornton Fry was a member as well.

Weaver shared Shannon's small-town roots and love of working with his hands. When he wasn't off practicing science or funding it, he was at home, "chopping wood, moving rocks, gardening, puttering in his shop." A shy and introspective boy, Weaver discovered a passion for engineering inside a small dry-cell motor, a Christmas present he promptly disassembled:

> I didn't know any name to apply to this sort of activity—I didn't know (or care, I suspect) whether anyone could earn his living doing this kind of thing. But it was perfectly clear to me that taking things apart and finding out how they are constructed and how they work was exciting, stimulating, and tremendous fun. It may well be the case that in the small rural village where I lived . . . there was not a single person who had any real concept of what the word "science" meant. I was accordingly told that this was "engineering"; and from that time until I was a junior in college, I assumed without question that I wanted to be an engineer.

It was a reflection that could have just as easily come from the pen of his grantee, Claude Shannon. But there the similarities stopped. Where Shannon was an avowed atheist, Weaver was devout—and saw science as self-evident proof of the divine. "I think that God has revealed Himself to many at many times and in many places," Weaver wrote. "Indeed, He continuously reveals Himself to man today: every new discovery of science is a further 'revelation' of the order which God has built into His universe." Where Shannon was allergic to administrative work and bureaucracies of almost every kind, Weaver thrived in them. Where Shannon considered teaching a nettlesome requirement of university employment, Weaver relished it. And where Shannon could continue to pound away at a mathematical problem or research question until he struck sparks, pursuing problems with breathtaking intuition and instinct, Weaver had discovered that he possessed no such gift. In a remarkably self-aware reflection on his strengths and weaknesses, Weaver observed: "I had a good capacity for assimilating information, something of a knack for organizing, an ability to work with people, a zest for exposition, an enthusiasm that helped to advance my ideas. But I lacked that strange and wonderful creative spark that makes a good researcher. Thus I realized that there was a definite ceiling on my possibilities as a mathematics professor."

Below that ceiling, however, Weaver was a heterodox thinker whose passions ran the gamut; he published or worked on problems in engineering, mathematics, machine learning, translation, biology, the natural sciences, and probabilities. But unlike many of his colleagues, he believed in a world outside the confines of science and math, and he avoided the all-too-common insularity of the fields he worked in and the thinkers who worked in them. "Do not overestimate science, do not think that science is all that there is," he urged students in a 1966 talk. "Do not concentrate so completely on science that there's nobody in this room who is going to spend the next seven days without reading some poetry. I hope that there's nobody in this room that's going to spend the next seven days without listening to some music, some good music, some modern music, some music."

He lived by his words: he was, according to Fry, epicure enough to identify a sip of wine by its varietal, vineyard, and even vintage year. And he maintained a lifelong passion for Lewis Carroll's *Alice's Adventures in Wonderland*. By the mid-1960s he had collected 160 versions of the text in forty-two languages, a compilation that moved him to write *Alice in Many Tongues,* a study of the effect of the act of translation on the meaning of the tale.

Weaver was many things Shannon was not: a populist, a philosophizer, a human interface between science and the wider world. But at that precise moment, those differences mattered little: he was someone who saw the potential in Shannon and had the ability to put that potential to work on wartime projects. He awarded Shannon $3,000 and a ten-month contract, for a project called "Mathematical Studies Relating to Fire Control." Shannon would complete much of his work in this field while remaining in Princeton, but he collaborated with two Bell Labs engineers, Ralph Blackman and Hendrik Bode, who would join the impressive group of influences and mentors in Shannon's life.

Fire control was, essentially, the study of hitting moving targets. The targets were anything and everything the enemy could hurl through the air to cause damage—planes, rockets, ballistics. Imagine a gun firing a single shot at a target. Now imagine that the gun is the size of a two-story house, that it is placed on a moving Navy ship in the middle of the ocean, and that it is trying to shoot down an enemy fighter moving at 350 miles per hour. That's a rough description of the challenge of fire control, and it was put to the mathematics group at Bell Labs, among others, to design the machines that could successfully solve that problem. Accurately determining the vertical coordinates, the horizontal coordinates, the speed of the ballistics, the likely position of the target, the time from launch to impact—all of this had to be processed by machine, without error, under fire, in split seconds.

The earliest days of the war revealed how badly the Allied defense systems needed an upgrade. The German Luftwaffe, an air force disbanded by the Treaty of Versailles, had been impressively reconstituted under Hitler and Hermann Göring. It had been responsible for the destruction of Guernica during the Spanish Civil War, and for the London Blitz, and as the war dragged on, the German military developed and deployed some of the world's first cruise and ballistic missiles.

Still, what special insight did telephone engineers have into threats like that? A lot, as it turned out. "At first thought it may seem curious that it was a Bell Telephone Laboratories group which came forward with new ideas and techniques to apply to the AA [anti-aircraft] problems," Warren Weaver admitted. And yet, Weaver continued, the Bell Labs group was a natural fit, for two reasons. "First, this group not only had long and expert experience with a wide variety of electrical techniques. Second, there are surprisingly close and valid analogies between the fire control prediction problem and certain basic problems in communications engineering."

At the most basic level, the speed and quality of information was vital both to phone systems and fire control systems. A phone call reaching its intended recipient was a struggle against noise. An anti-aircraft missile hitting its target presented the same conceptual challenge: How to get from point A to point B with a minimum of interference? In the case of the missile, how to guard against the wind, or factor in the movement of the target, among a dozen other variables? Because both problems demanded the quick calculation of probabilities—the probable structure of a message, or the probable location of the target at any given moment—both required high-level statistical inference.

Both presented the challenge of building machines to accurately translate that math into action.

Of course, the Bell engineers tasked with working on the problem were under no illusions: even if the technological problems shared some qualities, the stakes couldn't have been more different. A fraction of a second of difference in the control of an anti-aircraft gun meant the difference between life and death. For Shannon, in particular, the fire control work represented some of the most concrete work he had done to date. Unlike, say, his research on genetics, there was nothing abstract about shooting down airplanes.

There were mechanical as well as conceptual similarities between communications and fire control work. Bell Labs had begun its work on fire control when one of its engineers discovered that an existing piece of communications technology—a potentiometer—could be repurposed as part of an anti-aircraft gun. The potentiometer was used as a sort of moving hinge that responded to variations of voltage in, say, a telephone or radio receiver. A young engineer at the Labs, David Parkinson, had also experimented with connecting a potentiometer to a pen on graph paper, which allowed it to plot the outputs of electromechanical systems. The idea that such a thing could help shoot down an aircraft had come to him, of all possible places, in a dream. In the dream,

> I found myself in a gun pit with an anti-aircraft gun crew. . . . There was a gun there which looked to me like—I had never had any close association with anti-aircraft guns, but possess some general information on artillery—like a 3 inch. It was firing occasionally and the impressive thing was that *every shot brought down an airplane*! After three or four shots, one of the men in the crew smiled at me and beckoned me to come closer to the gun. When I drew near, he pointed to the exposed end of the left trunnion. Mounted there was the control potentiometer of my level recorder! There was no mistaking it—it was the identical item.

Reflecting on the dream the next morning, he realized that "if my potentiometer could control the pen on the recorder, something similar could, with suitable engineering, control an anti-aircraft gun." Parkinson took the idea to his superiors. They ran it up the chain at Bell, and it went from there to the Army Signal Corps. The T-10 director, built a few years later, was the culmination of Parkinson's dream, a project that drew on the Labs' years of communications work. In building it, they borrowed not only the language of radio and telephony, with which they were most familiar, but the component parts, too. Later renamed the M-9, the device would see combat: more than 1,500 M-9s were produced and put into the field. With M-9s directing the guns, the number of shells required to hit the average enemy aircraft was cut from thousands to just 100.

Many hands made the anti-aircraft directors a reality; Shannon's were two of them. "I think England would've been entirely demolished if they hadn't had these directors," he said after the war. While manned planes could still, with a little luck, evade anti-aircraft fire, "the buzzbombs and V1 missiles were going in perfectly straight lines and at moderate enough speed and everything, so that they could be predicted very well by these anti-aircraft directors, and they knocked down like 95% of them before they got to England, and I think if they hadn't had those England would've lost."

Shannon's particular contribution focused on the problem of "smoothing." The earliest prototypes of the gun directors produced occasionally erroneous readouts, leading to jerkiness in the motions of the guns. Smoothing was the process of cleaning up that data, without adding any delay in the calculation. The result of Shannon's work, five technical papers of varying lengths, was twofold: an upgrade to the original T-10 model, and a later report on the statistics of smoothing in general. The former never saw the light of day; the latter became a key work in the field.

What might Shannon have taken away from all of this? David Mindell, a historian of technology, put it like this:

> The wartime efforts of Bell Labs in fire control contributed to a new vision of technology, a vision that treated different types of

> machinery (radar, amplifiers, electric motors, computers) in analytically similar terms—paving the way for information theory, systems engineering, and classical control theory. These efforts produced not only new weapons but also a vision of signals and systems. Through ideas and through people, this vision diffused into engineering culture and solidified as the technical and conceptual foundations of the information age.

In other words, the research may have paid immediate dividends, but the real source of value was the analogy. Scientific breakthrough by way of analogy has a rich history. It's said that Galileo's work on pendulums began in a church in Pisa, where he stared at a lamp swinging through the air and timed it with his pulse. Newton, of course, had his apple. Einstein imagined himself chasing a beam of light. As for Shannon: Wasn't the work of tracking the evasive-yet-predictable path of an aircraft a rigorous training course in thinking probabilistically? If the position of the aircraft was best understood in that way—not in terms of where the target was, but in terms of where it was most likely to be—what other elusive objects might be targeted in the same way?

In Shannon's report on the topic, cowritten with two other Labs researchers, they acknowledge that the problem is a "special case of the transmission, manipulation, and utilization of intelligence. . . . The input data . . . are thought of constituting a series in time similar to weather records, stock market prices, production statistics, and the like." That thought presaged a key insight of Shannon's later work: that sources of "intelligence" as disparate as the trajectory of a missile and the output of a stock ticker, the pulses in a telegraph line and the instructions in a cell nucleus, had something heretofore unsuspected in common.

Those insights were still years away. What mattered in the here and now for Shannon was that his work for the NDRC had impressed the higher-ups. "He did some stunning work for us," Weaver later said. Fry, who had first seen Shannon in action during the summer of 1940, now had ample evidence of Shannon's promise. It didn't take long for him to extend a full-time offer to join Bell Laboratories as a research mathematician.

For Shannon, the offer must have come as a relief—not just professionally, but personally. The accounts of that time depict a man on edge—and understandably so. The pressure of the war combined with the collapse of his marriage had left Shannon shattered. "For a time it looked as though he might completely crack up nervously and emotionally," Weaver remembered. "It is Thornton Fry who deserves the primary credit for getting him out of that state, and for offering him work in the Bell Laboratories. The rest of it is history."

10

A Six-Day Workweek

This has not been a scientist's war; it has been a war in which all have had a part. The scientists, burying their old professional competition in the demand of a common cause, have shared greatly and learned much.

—Vannevar Bush

The Bell Labs headquarters in Manhattan's West Village were a scientific smorgasbord: chemical labs, vast production rooms, and "a warren of testing labs for phones, cables, switches, cords, coils, and a nearly uncountable assortment of other essential parts." And now, with a host of new wartime projects under way and hundreds of new faces streaming through the office, including many in military uniforms, the thirteen stories on the Hudson's edge felt especially chaotic. Even as several hundred Labs employees departed for active-duty service in the wake of Pearl Harbor, Bell's in-house workforce swelled: 4,600 employees became over 9,000 in only a matter of a few years. More than 1,000 research projects were launched, each one a small piece of the war machine. The tempo picked up accordingly; "a six-day workweek became the norm," Gertner writes.

Bell Labs wasn't alone in feeling the pressures of the war. Conflict overseas placed crushing new demands on much of the nation's scientific elite and the institutions that housed them. As Fred Kaplan explained in his history of wartime science, "It was a war in which the talents of scientists were exploited to an unprecedented, almost extravagant degree." There were urgent questions that needed answers, and the scientifically literate were uniquely equipped to answer them. Kaplan cataloged just a few:

> How many tons of explosive force must a bomb release to create a certain amount of damage to certain types of targets? In what sorts of formation should bombers fly? Should an airplane be heavily armored or should it be stripped of defenses so it can fly faster? At what depths should an anti-submarine weapon dropped from an airplane explode? How many anti-aircraft guns should be placed around a critical target? In short, precisely how should these new weapons be used to produce the greatest military payoff?

A generation of physicists and mathematicians was unleashed on puzzles like these.

One of the most insightful surveys of wartime mathematics comes from J. Barkley Rosser, a University of Wisconsin professor who interviewed some 200 mathematicians who, like him, had been pressed into national service. Rosser concluded that mathematicians acted as a kind of accelerant, helpful in speeding up research and development that would otherwise have been painfully manual and slow.

> The attitude of many with the problems they were asked to solve was that the given problem was not *really* mathematics but, since an answer was needed, urgently and quickly, they got on with it. . . . Without a person with competence to supply an answer by mathematics, the person with the problem would have had to resort to some scheme of experimental trial and error. This could be very expensive. Worse still, it could be very time-consuming, and everybody wished to get the War over as quickly as possible. So though mathematicians turned up their noses at most of the problems brought to them, they did so privately, and labored enthusiastically to produce answers.

And so hundreds of the world's top mathematical minds put their personal research aside, swallowed various degrees of pride, and gathered at the outposts of Los Alamos, Bletchley Park, Tuxedo Park—and Bell Labs, where wartime contracts brought a fresh-from-fellowship Claude Shannon into contact with the latest in military technology and thought.

For men like Vannevar Bush, James Conant, John von Neumann, J. Robert Oppenheimer, and others, the war lifted the veil on their work. They were invited into the councils of power, asked to advise presidents, and tasked with steering millions of dollars of men and matériel. Many of these men had made modest names for themselves in the worlds of science and engineering, but in the arena of wartime politics, their work would receive wide public recognition.

Shannon, too, might have entered this elite group—but he chose not to. "He couldn't care less about what was happening in Europe," his girlfriend at the time remembered. Unlike many of his contemporaries, Shannon displayed no ambitions for the high-wire world of government. He made no special effort to earn assignments related to the war effort, nor did he go out of his way to play up his fire control research. This wasn't, as it might have been for some of his less-sought-after contemporaries, for lack of access. With Vannevar Bush as a trusted mentor and a resume fat with fellowships and prestigious institutions, Shannon could have navigated his way to the high government post of his choosing.

But he didn't. If anything, his reaction to the war work was quite the opposite: the whole atmosphere left a bitter taste. The secrecy, the intensity, the drudgery, the obligatory teamwork—all of it seems to have gotten to him in a deeply personal way. Indeed, one of the few accounts available to us, from Claude's girlfriend, suggests that he found himself largely bored and frustrated by wartime projects, and that the only outlet for his private research came on his own time, late at night. "He said he hated it, and then he felt very guilty about being tired out in the morning and getting there very late. . . . I took him by the hand

and sometimes I walked him to work—that made him feel better." It's telling that Shannon was reluctant, even decades later, to talk about this period in any kind of depth, even to family and friends. In a later interview, he would simply say, with a touch of disappointment in his words, that "those were busy times during the war and immediately afterwards and [my research] was not considered first priority work." This was true, it appears, even at Bell Labs, famously open-minded though it may have been.

There was something else, too: as Rosser suggested, the math problems brought forth by the war were hardly math at all—or, at least, they were beneath anyone considered worth working on them. The defense establishment had, in a sense, overinvested in brainpower. In Rosser's words, one of his colleagues

> insisted to his dying day . . . that he never did an iota of mathematics during the War. True enough, the problems were mostly very pedestrian stuff, as mathematics. I was never required to appeal to the Gödel incompleteness theorem, or use the ergodic theorem, or any other key results in that league. One time the tedium was relieved when I had to do something with orthogonal polynomials, and I was glad to get out the Szegő tome and bone up a bit. But mostly I was working out how fast our rockets would go, and where. On a good day, some problem would be up to the level of a junior course in mathematics.

Call it an extreme case of mathematical snobbery, but we can imagine Shannon sharing the sentiment, even if he wasn't willing to write it down for posterity. Shannon, fresh from the tony confines of Princeton and MIT and the exciting problems that had consumed his young career, may well have considered it a step backward to calculate where, when, and how large airborne objects went boom.

Yet in his fortunate life, surely one of Shannon's great strokes of luck was that he found his way to full-time work at Bell Labs not long before the formal American entry into the war. Although he couldn't have known it then, his military research would prove to be more than just a way of avoiding combat. His principal projects—secrecy systems and cryptography—would introduce him to what cutting-edge computer technology could achieve. Even if he came to all of it reluctantly, he was exposed to it nonetheless. Only later would he hint that it was in the war's technologies that he began to see the broad outlines of the technological progress to come—progress that he, in his own way, would help bring to pass.

11

The Unspeakable System

Cryptography was the war's white noise: it was ubiquitous, and yet only those paying the closest attention could pick it up. It was one of the least understood components of the war machine. Compared to, say, the nuclear bomb, a visible and white-hot expression of the power of physics, the products of cryptographic analysts were arcane and mysterious—and kept classified for a generation or more.

From the war's earliest days, the challenge of sending and receiving coded messages—and of cracking the enemy's messages—drew some of the world's finest minds in mathematics, science, and computing. The technology called forth by the need to break codes was among the war's great triumphs. The cottage industry of computer code names—ENIGMA, ENIAC, MANIAC, TUNNY, BOMBE, COLOSSUS, SIGSALY, and so on—evinced a demand for codebreaking power that was spurring a revolution in computing, presided over by secretive intelligence bureaucracies.

But this deeply bureaucratic story is not how the tale of wartime codebreaking is often told. Cryptography is made to seem like the work of brilliant lone wolves, scribbling away in solitude. As Colin Burke puts it in a National Security Agency history of cryptologic efforts called *It Wasn't All Magic,* "Such an image of a heroic cryptanalysis is far from being true or useful. Cryptanalytic and technological victories have not come as easily as that. . . . Typical cryptanalysis was and remains a continuing struggle to discover patterns and to make sense out of mountains of raw data." Because the cryptographic bureaucracies grew in secret, and because many of their files still remain classified—Burke's own history, completed in 1994, was only declassified in 2013—the actual substance of their work wasn't (and still isn't) well understood by the public.

But it was always there, the dull background hum of the war: codes broken, codes built, thousands of conversations deciphered, reams of data and text sorted by hand and machine. The war for signals intelligence was as much about code-making as it was about codebreaking, as illustrated by one famous and tragic story. On the morning of December 7, 1941, George Marshall, the Army's chief of staff, had an important message to send to his Pacific Command: Japan had decided that it could no longer mediate its differences with the United States through politics; war was likely. But how to transmit this information? The only system available to the nation's top military and political leaders had long been considered insecure. The message was sent instead by the comparatively slower radio telegraph. Tragically, it arrived after the attack on Pearl Harbor was over. The near destruction of the Pacific Fleet was, among much else, a wake-up call to America's code writers.

The Axis powers, too, set their finest minds and technologies to work trying to intercept and decode enemy conversations. Walter Schellenberg, the head of the Third Reich's Foreign Intelligence Service, details one such success late in the war:

> Early in 1944 we hit a bull's eye by tapping a telephone conversation between Roosevelt and Churchill which was overheard and deciphered by the giant German listening post in Holland. Though the conversation was scrambled, we unscrambled it by means of a highly complicated apparatus. It lasted almost five minutes, and disclosed a crescendo of military activity in Britain,

thereby corroborating the many reports of impending invasion. Had the two statesmen known that the enemy was listening to their conversation, Roosevelt would hardly have been likely to say good-bye to Churchill with the words, "Well, we will do our best—now I will go fishing."

Cryptology represents a problem of both software and hardware. The "software" can, in principle, be anything. In one well-known example, about 500 Navajo Indians were recruited in World War II to transmit coded messages because their native tongue was complex enough—and unfamiliar enough—to evade detection by the Axis powers. That's the essence of cryptology: a series of substitutions, trades of one letter or word for another letter or word (or language). Technology can help augment a code's difficulty by ramping up the complexity of the substitutions. And thus, advances in cryptographic hardware rendered World War II–era codes exponentially more complex. They enabled cryptographers, for instance, to easily encrypt each letter of a message with a different cipher alphabet, which would render the message as a whole far more resilient against codebreaking.

This was where Bell Labs entered the fight: what the country needed was computing power of the sort that would enable more efficient encryption of messages and speedier cracking of enemy codes. One example was "Project X"—the most ambitious speech-scrambling system of that era. It was initiated in the winter of 1940 and took on new urgency with the American entry into the war. The project, also known as the SIGSALY system, consisted of "some forty racks of vacuum tube–powered electrical equipment weighing about fifty-five tons, taking up 2,500 square feet and requiring 30,000 watts of power." According to one estimate, the system had a $5,000,000 budget in 1943, and it employed a platoon of thirty workers. So secret was the system's internal logic that the patents associated with it were not publicly divulged until 1976. To listen to a scrambled message over its wires was to hear something that sounded "rather like Rimsky-Korsakov's bravura violin spectacular 'The Flight of the Bumblebee.'" To critics who thought the output curious or difficult to understand, William Bennett, one of SIGSALY's engineers, had a prompt reply: "Accept distortion for security."

In one sense, SIGSALY looked like caricature of a mid-century computer: it occupied an entire room, demanded round-the-clock air conditioning, and produced small outputs for the enormous inputs it required. (This was an open joke: "Members working on the job occasionally remarked about the terrible conversion ratio—30 kW of power for 1 milliwatt of poor-quality speech.") But on the other hand, none of that mattered. Andrew Hodges, in his biography of Alan Turing, noted the obvious reason why: "It worked, which was the main thing. For the first time, secret speech could cross the Atlantic."

At the heart of the SIGSALY system was a technology known as the Vocoder. Its creator, later celebrated as an engineering genius, almost didn't become an engineer at all. Homer Dudley had ambitions of becoming a teacher. He had even given it a try, teaching fifth through eighth graders, and then high school students. While the material was no difficulty for someone of his intellect, he was never able to master the art of classroom discipline. Dudley discovered, like many a teacher before and since, that the real challenge lay in maintaining order among the prepubescent—and that he wasn't up to it. So he abandoned teaching to study electrical engineering, joining the technical department of Western Electric, the precursor of Bell Labs. It was a more auspicious career choice: over four decades, his work on telephony and speech synthesis resulted in thirty-seven patents.

Unbeknownst to him at the time, his most significant achievement would have global implications. Dudley hypothesized that the sounds of the human voice could be mimicked by machine: if the human voice was, at the most basic level, merely a series of vibrations in the air, there was no reason why those vibrations could not be mechanically reproduced. Setting out to test his premise, he built a pair of machines that did just that: one to encode speech electronically (the Vocoder, short for "Voice Encoder"), and one to reverse the process and output machine-generated speech (the Voder, or "Voice Operation Demonstrator"). Vannevar Bush was in the audience for the latter's debut at the 1939 World's Fair, where it proved a hit. As he recalled,

> At a recent world fair a machine called a Voder was shown. A girl stroked its keys and it emitted recognizable speech. No human vocal cords entered into the procedure at any point; the keys simply combined some electrically produced vibrations and passed these on to a loudspeaker. In the Bell Laboratories there is the converse of this machine, called a Vocoder. The loudspeaker is replaced by a microphone, which picks up sound. Speak to it, and the corresponding keys move.

Only later was Dudley's invention pressed into military service. In creating data out of the human voice, the Vocoder became a perfect building block for SIGSALY's engineers. One of the central challenges of a code-writing system is this: each new letter or word introduced into a message opens up a fresh possibility of detection by the enemy; the less that's communicated, in other words, the better. Because the Vocoder attempted to encode and reproduce vowels and consonants with as little energy consumed as possible, it wrung a great deal of the redundancy out of human speech; the result was an economy of information transmitted. In other words, what was encoded was only what was needed, nothing more, thus reducing the amount of information that would be liable to decryption by the enemy.

The challenge of communicating a maximum of information with a minimum risk of detection was deadly serious, one of the most urgent and complex wartime problems. Bell Labs was one of the national leaders in such work, even winning awards in the field, including "Best Signal Processing Technology" in 1946. SIGSALY remained classified, so none of its inner workings were disclosed to the awards ceremony audience. But the Labs' representative did accept the honor with an encrypted phone call: "Phrt fdygui jfsowria meeqm wuiosn jxolwps fuekswusjnvkci! Thank you!"

Shannon, for his part, was one of a team of almost thirty people working on the many pieces of the SIGSALY project. He was tasked with checking the algorithms that would allow the message to be suitably and securely reproduced on the receiving end. So secret was SIGSALY that, even as a member of the project team, Shannon was not told what all his number-crunching was for. But the work gave him a window into the world of encoded speech, transmission of information, and cryptography—a synthesis that, at that moment in history, may not have taken place anywhere other than at Bell Labs. As Shannon observed, "not a lot of laboratories had voice encoding devices for scrambling speech."

Shannon would observe later that cryptography was "a very down to earth discipline, how the cryptographer should go to work and what he should do." But much of Shannon's work wasn't designed for cryptographers. When Shannon worked on encryption outside the confines of the SIGSALY project, he was writing more for an audience of "mathematicians or philosophers of cryptography" than for hands-on code writers. As he himself admitted, his cryptography paper "didn't get as much good response . . . as I expected from the cryptographers." A later writer would admit that Shannon's paper on the topic of cryptography had "the feel of a what-can-I-contribute-to-the-war-effort undertaking of which there must have been many."

As in other areas of Shannon's life, his most important work in cryptography yielded a rigorous, theoretical underpinning for many of a field's key concepts. Shannon's exposure to day-to-day cryptographic work during the war, it seems, was important—but its primary purpose was as grist for a paper that would only be published in classified form on September 1, 1945—one day before the Japanese surrender was signed. This paper, "A Mathematical Theory of Cryptography—Case 20878," contained important antecedents of Shannon's later work—but it also provided the first-ever proof of a critical concept in cryptology: the "one-time pad."

The one-time pad system was the conceptual basis of Bell Labs' Vocoder, though it was first devised as early as 1882. It requires that a coded message be preceded by the key to decode it, that the key be a secret, entirely random set of symbols the same size as the message, and that the key be used only once. It took Claude Shannon, and the space of more than a half century, to prove that a code constructed under these stringent (and usually impracticable) conditions would be entirely unbreakable—that perfect secrecy within a cryptographic system was, at least in theory, possible. Even with unlimited computing power, the enemy could never crack a code built on such a foundation.

Shannon's cryptographic work was released into the dark of the intelligence apparatus, a world of classifications and secrecy where the work's reception was concealed even from its author. Of the people in this world, Shannon would say: "They were not a very talkative bunch, you could say that. They were the most secretive bunch of people in the world. It's very hard to even find out for example who are the important cryptographers in this country." The paper would not be available to a wider audience for another five years. The work's true import, in the end, was not in the creation of an invincible code, but rather in the way that its suppressed insights ultimately resurfaced, at the heart of Shannon's revolutionary theory of information: "It was a great flow of ideas from one to the other, back and forth."

12

Turing

Shannon's cryptography work had one other lasting import: it brought him into contact with another giant of the digital age, Alan Turing. In 1942, Turing came to America as a part of a government-initiated tour of military encryption projects. By this point, his reputation preceded him. He had demonstrated astonishing mathematical precocity in primary school; mastered Einstein's work by sixteen; and by twenty-three had been elected a fellow of King's College, Cambridge. In 1936, he dreamed up the Turing Machine, a landmark thought experiment that served as the theoretical underpinning of the modern computer.

Turing had also begun the codebreaking that would make him a world-historic figure. It was cryptography that had brought him to the United States, on orders to connect with American counterparts, meet the military leaders in the space, and inspect the American machinery for quality and security. This extended to, among other projects, SIGSALY; if British leaders were to be on the receiving end of its buzzing, Dr. Turing would have to give his stamp of approval that the system was unbreakable.

The secrecy of the subject matter, the reputations of Turing and Shannon, and the atmosphere of the war has lent this meeting of the minds an air of intrigue and mystery. But there was nothing cloak-and-dagger about their interactions. According to Turing's biographer, Andrew Hodges, Shannon and Turing met daily over tea, in public, in the conspicuously modest Bell Labs cafeteria. Turing was envious, in a way, of Shannon's multifaceted career: "Here [Turing] met a person who had been able to take the part of an academic, philosophical engineer, the role that Alan might have liked had the English system allowed for it." Shannon, for his part, was amazed by the quality of Turing's thinking. "I think Turing had a great mind, a very great mind," Shannon later said.

Neither left a record of their conversations, but we do know the one topic they avoided. "We talked not at all about cryptography. . . . I don't think we exchanged word one about cryptography," Shannon explained. Asked whether or not he knew what Turing was working on, Shannon responded that he only knew it in broad strokes. "Certainly not the nitty gritty. I knew or surmised what he was doing. . . . I had no concept of the Enigma machine. . . . I didn't know of that nor that he was a crucial figure in it." The interviewer pressed him, asking him why Shannon, with his own passion for and experience in cryptographic puzzles, wouldn't have probed Turing further. Shannon's response was simple and to the point: "Well, in the wartime you didn't ask too many questions."

To some, Shannon and Turing's professed ignorance of the other's work might seem like a pair of codebreakers smartly covering the tracks of their relationship. But it's entirely plausible that neither man wished to put the other in an uncomfortable or compromised position. The work on both sides was highly classified. The information they had each been given was, if not for their eyes only, certainly for those of a limited audience. It's no surprise that, given a break from their labors over tea and cake, the two men would spend time talking about anything other than the work that occupied their days.

Another reason: questions about how much the Americans were willing to share with the British were still up in the air. Even for someone of Turing's stature and reputation, obtaining the relevant clearances for his American trip had been an ordeal. He had, amazingly enough, been detained by the American authorities on arrival: "I reached New York on Friday November 12th. I was all but kept on Ellis Island by the Immigration Authorities who were very snooty about my carrying no orders and no evidence to connect me with the F.O. [Foreign Office]."

Months earlier, the American general Rex Minckler had rejected Turing's application to enter Bell Laboratories. Though the application was, at length, finally approved, it only marked the beginning of a protracted tussle with the Bell Labs and national security bureaucracies. Turing wrote:

> I had been intending to report to Potter at Bell Laboratories without any more formality than a preliminary phone call. This was apparently all wrong. . . . There was some trouble because no arrangements for me to see anything other than unscrambling projects had been confirmed in writing, whereas I had come out on the understanding that I was to see everything there was in the way of speech secrecy work. . . . [I] immediately came up against a veto on any British people visiting anything at all in the speech scrambling line. Captain Hastings then took a hand, and brought the pressure to bear on General Colton, and all now seems to be well.

Turing wasn't alone in his aggravation with the Americans, nor did that frustration limit itself to immigration and security clearance snafus. Though the Allied forces had been working together in the war effort since the Lend-Lease Act, they didn't always see eye to eye on cryptographic matters.

Competing systems, methods, and personalities left the two sides perpetually suspicious of one another; tempers ran short, and egos bruised easily. The conflict was due, in part, to essential differences between the American and British war machines, and to their "incomplete alliance." For all their ramp-up in industrial production of military goods, the British simply could not keep pace with the Americans in scale or speed. Turing saw this firsthand, and his respect for American brains was, in some ways, the inverse of his respect for American brawn. After a visit to the Navy Department, for instance, he wrote:

"I am persuaded that one cannot very well trust these people where a matter of judgment in cryptography is concerned." Still, "I think we can make quite a lot of use of their machinery."

That attitude of qualified mistrust was common on both sides. Certain British successes in codebreaking, for instance, were kept hidden from the Americans, and Turing did not have a clear sense of what he was and was not allowed to share with his hosts. Already known for his brusque, off-putting manner, Turing wasn't going to ingratiate himself with the Americans when the very purpose of his mission was so clouded with uncertainty, nor was he the kind of natural diplomat capable of finessing the problem.

But in a way, the bad feelings between the two sides, and the muzzles placed on both Shannon and Turing, freed the pair to talk about their other common interests. A friendship blossomed where only a modest professional relationship might have sprung up had they been freer to talk about cryptographic matters. Even before the war years, both Turing and Shannon had shared similar extracurricular enthusiasms, and both had been dancing around the same set of cutting-edge ideas. Over their cups of tea, remembered Shannon, "we would talk about mathematical subjects." In particular, thinking machines were much on the two men's minds—in Shannon's words, "the notion of building computers that will think and what you can do with computers and all that." He went on:

> Turing and I had an awful lot in common, and we would talk about that kind of question. He had already written his famous paper about Turing Machines, so called, as they call them now, Turing Machines. They didn't call them that then. And we spent much time discussing the concepts of what's in the human brain. How the brain is built, how it works and what can be done with machines and whether you can do anything with machines that you can do with the human brain and so on. And that kind of thing. And I had talked to him several times about my notions on Information Theory, I know, and he was interested in those.

They were, both, taken by the promise of early computing and intrigued by the idea of chess-playing computers. Here's Shannon in 1977:

> Well, back in '42 . . . computers were just emerging, so to speak. They had things like the ENIAC down at University of Pennsylvania. . . . Now they were slow, they were very cumbersome and huge and all, there were computers that would fill a couple rooms this size and they would have about the ability of one of the little calculators that you can buy now for $10. But nevertheless we could see the potential of this, the thing that happened here if things ever got cheaper and we could ever make the up-time better, sort of keep the machines working for more than ten minutes, things like that. It was really very exciting.
>
> We had dreams, Turing and I used to talk about the possibility of simulating entirely the human brain, could we really get a computer which would be the equivalent of the human brain or even a lot better? And it seemed easier then than it does now maybe. We both thought that this should be possible in not very long, in ten or 15 years. Such was not the case, it hasn't been done in thirty years.

At his core, Shannon was a private person, with fewer confidants than his stature within the scientific community might suggest. In a life of exposure to many of the world's leading scientists, mathematicians, and thinkers, Shannon always gave the impression of a wallflower, someone who may have found himself in the same conference halls as many of the luminaries of his day but generally waited for the conversation to come to him. He wasn't a steady correspondent with the major figures he knew, attended only a fraction of the conferences to which he was invited, and was the sort of person for whom the concept of "networking" was distasteful when applied to anything other than telephone lines. All of which is to say that the fact he connected as enthusiastically as he did with Turing is as remarkable as anything the pair discussed. That Shannon could, in the brief few months they had together at Bell Labs, win Turing's confidence and friendship says a great deal about each man's high opinion of the other. Turing was, in Shannon's words, "a very, very impressive guy." Turing even visited Shannon at home, a rarity for a host who so preferred his own company—and a guest who did, too.

After Turing's return to Great Britain, he and Shannon met one final time, in the war's aftermath. In 1950, Claude traveled to London for a conference and took time to visit Turing and his laboratory. As Shannon remembered:

> While there we went over to Turing's Laboratory in Manchester at the University of Manchester. . . . He was interested in programming a computer for chess, which . . . was a problem that had interested me a great deal. And he was working away with programming a computer at this time. And he had an office up there, and there was a computer downstairs. This was in the early days of computers.

The two discussed Turing's efforts at programming. Even decades later, Shannon would recall one of Turing's inventions:

> So I asked him what he was doing. And he said he was trying to find a way to get better feedback from a computer so he would know what was going on inside the computer. And he'd invented this wonderful command. See, in those days they were working with individual commands. And the idea was to discover good commands.
>
> And I said, what is the command? And he said, the command is put a pulse to the hooter, put a pulse to the hooter. Now let me translate that. A hooter . . . in England is a loudspeaker. And by putting a pulse to it, it would just be put a pulse to a hooter.
>
> Now what good is this crazy command? Well, the good of this command is that if you're in a loop you can have this command in that loop and every time it goes around the loop it will put a pulse in and you will hear a frequency equal to how long it takes to go around that loop. And then you can put another one in some bigger loop and so on. And so you'll hear all of this

> coming on and you'll hear this "boo boo boo boo boo boo," and his concept was that you would soon learn to listen to that and know whether when it got hung up in a loop or something else or what it was doing all this time, which he'd never been able to tell before.

It was an altogether enjoyable visit, a postwar reunion for two of the founders of the Information Age. It was also the last time that they would speak in person. Four years after Shannon's visit, Turing—following a conviction for "gross indecency" at a time when homosexuality was criminalized—died of cyanide poisoning. His death was ruled a suicide, though its circumstances remain in dispute to this day.

13

Manhattan

The ancient art of mathematics . . . does not reward speed so much as patience, cunning and, perhaps most surprising of all, the sort of gift for collaboration and improvisation that characterizes the best jazz musicians.

—**Gareth Cook**

With his marriage to Norma over, Shannon was a bachelor again, with no attachments, a small Greenwich Village apartment, and a demanding job. His evenings were mostly his own, and if there's a moment in Shannon's life when he was at his most freewheeling, this was it. He kept odd hours, played music too loud, and relished the New York jazz scene. He went out late for raucous dinners and dropped by the chess clubs in Washington Square Park. He rode the A train up to Harlem to dance the jitterbug and take in shows at the Apollo. He went swimming at a pool in the Village and played tennis at the courts along the Hudson River's edge. Once, he tripped over the tennis net, fell hard, and had to be stitched up.

His home, on the third floor of 51 West Eleventh Street, was a small New York studio. "There was a bedroom on the way to the bathroom. It was old. It was a boardinghouse . . . it was quite romantic," recalled Maria Moulton, the downstairs neighbor. Perhaps somewhat predictably, Shannon's space was a mess: dusty, disorganized, with the guts of a large music player he had taken apart strewn about on the center table. "In the winter it was cold, so he took an old piano he had and chopped it up and put it in the fireplace to get some heat." His fridge was mostly empty, his record player and clarinet among the only prized possessions in the otherwise spartan space. Claude's apartment faced the street. The same apartment building housed Claude Levi-Strauss, the great anthropologist. Later, Levi-Strauss would find that his work was influenced by the work of his former neighbor, though the two rarely interacted while under the same roof.

Though the building's live-in super and housekeeper, Freddy, thought Shannon morose and a bit of a loner, Shannon did befriend and date his neighbor Maria. They met when the high volume of his music finally forced her to knock on his door; a friendship, and a romantic relationship, blossomed from her complaint.

Maria encouraged him to dress up and hit the town. "Now this is good!" he would exclaim when a familiar tune hit the radio on their drives. He read to her from James Joyce and T. S. Eliot, the latter his favorite author. He was, she remembered, preoccupied with the math problems he worked over in the evenings, and he was prone to writing down stray equations on napkins at restaurants in the middle of meals. He had few strong opinions about the war or politics, but many about this or that jazz musician. "He would find these common denominators between the musicians he liked and the ones I liked," she remembered. He had become interested in William Sheldon's theories about body types and their accompanying personalities, and he looked to Sheldon to understand his own rail-thin (in Sheldon's term, ectomorphic) frame.

A few Bell Labs colleagues became Shannon's closest friends. One was Barney Oliver. Tall, with an easy smile and manner, he enjoyed scotch and storytelling. Oliver's easygoing nature concealed an intense intellect: "Barney was an intellect in the genius range, with a purported IQ of 180," recalled one colleague. His interests spanned heaven and earth—literally. In time, he would become one of the leaders of the movement in the search for extraterrestrial life. Tom Perkins, cofounder of the famed Kleiner Perkins venture capital firm, remembered Oliver's ability to seize on a topic, no matter how obscure. "If the prospect of building devices to communicate with dolphins captured his fancy, that's what he did for months on end,"

Perkins recalled. He was the brains behind "Project Cyclops," the "ingenious and noble albeit unfulfilled" plan to connect 1,000 100-meter satellite dishes across a thirty-six-square-mile stretch of land with the goal of amplifying radio waves enough to detect interstellar chatter.

Oliver's earthbound pursuits were equally ambitious. They included "the world's first programmable desktop calculator," its handheld offspring, and the first Hewlett-Packard computer. Oliver also held the distinction of being one of the few to hear about Shannon's ideas before they ever saw the light of day. As he proudly recalled later, "We became friends and so I was the mid-wife for a lot of his theories. He would bounce them off me, you know, and so I understood information theory before it was ever published." That might have been a mild boast on Oliver's part, but given the few people Shannon let into even the periphery of his thinking, it was notable that Shannon talked with him about work at all.

John Pierce was another of the Bell Labs friends whose company Shannon shared in the off hours. At the Labs, Pierce "had developed a wide circle of devoted admirers, charmed by his wit and his lively mind." He was Shannon's mirror image in his thin figure and height—and in his tendency to become quickly bored of anything that didn't intensely hold his interest. This extended to people. "It was quite common for Pierce to suddenly enter or leave a conversation or a meal halfway through," wrote Jon Gertner. It was the by-product of a blazing fast mind. Early in his education, Pierce had so impressed the professor of his engineering class that he was promoted, mid-course, from student to teacher. At the Labs, Pierce acquired a similar reputation. He was known for having a knack for invention that was on par with the Labs' best.

Shannon and Pierce were intellectual sparring partners in the way only two intellects of their kind could be. They traded ideas, wrote papers together, and shared countless books over the course of their tenures at Bell Labs. One story, from a speech Pierce gave, illustrates their collaborations:

> One day I was talking casually with Claude Shannon, and he described to me in a few words the system a worker outside of the Bell Laboratories had devised. I didn't pay much attention while

> he was talking, but something of what he had said stayed with me. Then, later in the day, I saw certain advantages of this new system. The next day I went to see Claude and told him that this was a fine idea. As I explained the advantages, he agreed, but he observed that the system I was describing wasn't the one he had told me about at all. I had invented a new system by listening carelessly and pursuing my own thoughts.

Pierce told Shannon on numerous occasions that "he should write up this or that idea." To which Shannon is said to have replied, with characteristic insouciance, "What does 'should' mean?"

Oliver, Pierce, and Shannon—a genius clique, each secure enough in his own intellect to find comfort in the company of the others. They shared a fascination with the emerging field of digital communication and cowrote a key paper explaining its advantages in accuracy and reliability. One contemporary remembered this about the three Bell Labs wunderkinds:

> It turns out that there were three certified geniuses at BTL [Bell Telephone Laboratories] at the same time, Claude Shannon of information theory fame, John Pierce, of communication satellite and traveling wave amplifier fame, and Barney. Apparently the three of those people were intellectually INSUFFERABLE. They were so bright and capable, and they cut an intellectual swath through that engineering community, that only a prestige lab like that could handle all three at once.

Other accounts suggest that Shannon might not have been so "insufferable" as he was impatient. His colleagues remembered him as friendly but removed. To Maria, he confessed a frustration with the more quotidian elements of life at the Labs. "I think it made him sick," she said. "I really do. That he had to do all that work while he was so interested in pursuing his own thing."

Partly, it seems, the distance between Shannon and his colleagues was a matter of sheer processing speed. In the words of Brockway McMillan, who occupied the office next door to Shannon's, "he had a certain type of impatience with the type of mathematical argument that was fairly common. He addressed problems differently from the way most people did, and the way most of his colleagues did. . . . It was clear that a lot of his argumentation was, let's say, faster than his colleagues could follow." What others saw as reticence, McMillan saw as a kind of ambient frustration: "He didn't have much patience with people who weren't as smart as he was."

It gave him the air of a man in a hurry, perhaps too much in a hurry to be collegial. He was "a very odd man in so many ways. . . . He was not an unfriendly person," observed David Slepian, another Labs colleague. Shannon's response to colleagues who could not keep pace was simply to forget about them. "He never argued his ideas. If people didn't believe in them, he ignored those people," McMillan told Gertner.

George Henry Lewes once observed that "genius is rarely able to give an account of its own processes." This seems to have been true of Shannon, who could neither explain himself to others, nor cared to. In his work life, he preferred solitude and kept his professional associations to a minimum. "He was terribly, terribly secretive," remembered Moulton. Robert Fano, a later collaborator of Shannon, said, "he was not someone who would listen to other people about what to work on." One mark of this, some observed, was how few of Shannon's papers were coauthored.

Shannon wouldn't have been the first genius with an inward-looking temperament, but even among the brains of Bell Labs, he was a man apart. "He wouldn't have been in any other department successfully. . . . You would knock on the door and he would talk to you, but otherwise, he kept to himself," McMillan said. Slepian would put his apartness still more colorfully: "My characterization of his smartness is that he would have been the world's best con man if he had taken a turn in that direction." ("He would have taken that as a big compliment," his daughter later said.)

There was something else, too, something that might have kept him at a remove from even his close colleagues: Shannon was moonlighting. On the evenings he was at home, Shannon was at work on a private project. It had begun to crystallize in his mind in his graduate school days. He would, at various points, suggest different dates of provenance. But whatever the date on which the idea first implanted itself in his mind, pen hadn't met paper in earnest until New York and 1941. Now this noodling was both a welcome distraction from work at Bell Labs and an outlet to the deep theoretical work he prized so much, and which the war threatened to foreclose. Reflecting on this time later, he remembered the flashes of intuition. The work wasn't linear; ideas came when they came. "These things sometimes . . . one night I remember I woke up in the middle of the night and I had an idea and I stayed up all night working on that."

To picture Shannon during this time is to see a thin man tapping a pencil against his knee at absurd hours. This isn't a man on a deadline; it's something more like a man obsessed with a private puzzle, one that is years in the cracking. "He would go quiet, very very quiet. But he didn't stop working on his napkins," said Maria. "Two or three days in a row. And then he would look up, and he'd say, 'Why are you so quiet?' "

Napkins decorate the table, strands of thought and stray sections of equations accumulate around him. He writes in neat script on lined paper, but the raw material is everywhere. Eight years like this—scribbling, refining, crossing out, staring into a thicket of equations, knowing that, at the end of all that effort, they may reveal nothing. There are breaks for music and cigarettes, and bleary-eyed walks to work in the morning, but mostly it's this ceaseless drilling. Back to the desk, where he senses, perhaps, that he is on to something significant, something even more fundamental than the master's thesis that made his name—but what?

2

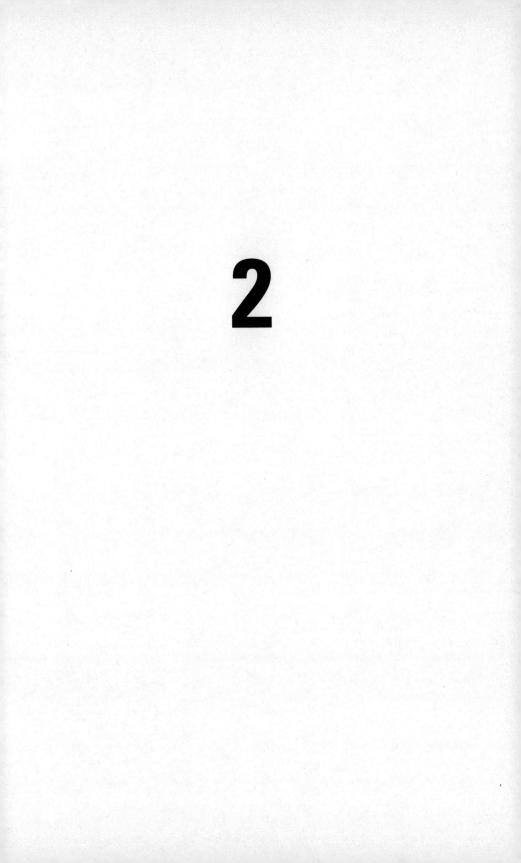

14

The Utter Dark

"Repeat, please."
"Please send slower for the present."
"How?"
"How do you receive?"
"Send slower."
"Please send slower."
"How do you receive?"
"Please say if you can read this."
"Can you read this?"
"Yes."
"How are signals?"
"Do you receive?"
"Please send something."
"Please send V's and B's."
"How are signals?"

Two and a half thousand tons of copper and iron had been strung two thousand miles across an ocean at the cost of millions of pounds and near shipwreck in order to act as conduit for the sputtering piece of failure you read above. That text represents an entire day's conversation across the great transatlantic telegraph cable, which, in the late summer of 1858, linked Europe to North America for twenty-eight days. The first message launched fireworks and knighthoods and euphoric editorials ("The Atlantic is dried up," announced the *Times* of London), but soon noise consumed the signal, and the wire went dead for hours at a time. Buried under three miles of water—sunk, in Kipling's words, "Down to the dark, to the utter dark, where the blind white sea-snakes are"—the cable was falling apart.

For the sake of those twenty-eight days of sporadic talk, British-American naval convoys had set out five separate times, unwinding the cable foot by foot as they steamed east across the Atlantic. On the fourth attempt, the ships were slammed by an historically ferocious storm. The British ship *Agamemnon,* a wooden vessel outfitted with both steam and sail, was caught in a weeklong gale, lurching from side to side by as many as 45 degrees, unbalanced by the tons of metal on her deck and in her hold—coils, wrote a nauseous newspaper correspondent on board, that resembled "nothing so much as a cargo of live eels." Four times the cable snapped. Only on the fifth voyage did it hold.

On each attempt, the most important passenger was a scientist who has already figured in this story: William Thomson, the future Lord Kelvin. His analog computer was two decades to come; in those days he was comparatively little known, and still without his Neptunian beard. But he was one of the world's leading experts on the wire transmission of information, though he didn't use that word. He had staked his reputation on the transatlantic project and been voted onto its board of directors as a scientific advisor; he served on each voyage, even the nearly fatal one, unpaid. An Australian reporter on board for the fifth attempt captured his mood when, in the middle of the night, electric current in the cable ceased and it looked to have snapped again: "The very thought of disaster seemed to overpower him. His hand shook so much that he could scarcely adjust his eye-glasses. The veins on his forehead were swollen. His face was deathly pale . . . yet in mind keen and collected, testing and waiting." But soon enough the signal came back to life, and Kelvin burst into laughter. A week later, the hills of County Kerry rose on the eastern horizon, and the cable was hauled onto the Irish shore, to be linked up with the European network.

A month later, it was inert junk on the bottom of the sea, destroyed by a disagreement.

Even before the Atlantic cable was laid, it was clear that messages sent through any underwater lines—across the English Channel, for instance—were especially prone to delay and distortion: transmitting a message through water is uniquely difficult. Because water, especially saltwater, is a natural conductor of electricity, a submerged cable is prone to find its electrical current leaching away. Compared to the signal borne by a dry cable, the signal sent along the length of a wet cable is far harder to discern.

No one understood this dilemma better than Thomson; it was, in large part, why he was aboard the *Agamemnon* to see the cable laid. Three years before the last sailing, his laboratory experiments in Glasgow had led him to argue that electrical transmission at a distance obeyed a "law of squares": the arrival time of a message increased with the square of the cable's length. Further, the signal's strength would grow increasingly attenuated the farther it had to travel. And if all of this were the case, then the only hope of reliable undersea communication was the thickest, best insulated—and most expensive—cable that had ever been constructed, paired with sensitive equipment to pick up faint signals at the far end.

But in 1858, in the absence of an ocean-spanning cable to put them to the test, those conclusions were very much in doubt. Strong financial incentives pushed the transatlantic project's backers to disregard Thomson: fortunes had begun to hang on the prospect of instantaneous communication across the ocean (imagine what a stockjobber in London could do with instant knowledge of commodity prices in Chicago), and Thomson's results came with the dispiriting warning that a truly reliable cable might cost more than it was worth. As misfortune would have it, Thomson's chief doubter was also his coworker: the head electrician of the transatlantic project.

Dr. O. E. Wildman Whitehouse was a retired surgeon and amateur electrical experimenter. That should not necessarily be counted as a slur on his expertise—the nineteenth century was a great age of gentleman amateurs. Against Thomson's university prestige, however, Whitehouse staked a frankly populist claim: he announced that the study of electricity and communication "is no longer the exclusive privilege of the philosopher." On the strength of his own flurry of experiments, he denounced the law of squares as "a fiction of the schools": a formula built for elegance in the pages of journals (it even *looked* like Newton's famous inverse-square law of gravity!), but one that fell apart in practice. Thomson struck the correct poses of Victorian decorum in response, but on his own copy of Whitehouse's work, he scribbled that it was "fallacious in almost every point." Where Thomson's results demanded a sturdier cable and finer signal detection, Whitehouse's called for brute force. As a later writer summed up his solution: "The further the electricity has to travel, the larger the kick it needs to send it on its way." To overcome distortion and delay, simply apply more power: it had the virtue of simplicity, and it underbid Thomson's plan, an inestimable advantage for a project that would live or die on the strength of the investments it could command.

In the end, it was a draw, and a farce. Thomson's "mirror galvanometer," his device for picking up faint electric signals, was installed at both ends, and disconnected by Whitehouse at every opportunity. The cable itself was built well below his standard of robustness. At the eastern end, on Valentia Island off the Irish coast, was Whitehouse—hooking up his massive, five-foot-long spark coils to push the signal through, and pumping electricity into the wire in 2,000-volt bursts.

Hauled in and out of ship holds, on and off decks, unwound and rewound, dropped to the seabed, snapped four times, spliced and respliced, the cable was already well punished by the time the first signal was sent. Now, subjected to Whitehouse's electrical barrage, its insulation fried and gave out in a matter of days. The last sorry message received at Valentia read: "Forty-eight words. Right. Right." Most of the messages sent and received on the celebrated wire were just like that: communications about communication, telegraphy as an especially bleak Samuel Beckett play.

Disobeying company orders, Whitehouse had a section of cable hauled up two miles offshore, searching for a wiring fault on which he could blame the breakdown; in the signal's last days, he was fired for insubordination. A postmortem parliamentary report made him the face of the failure (though scholars have argued more recently that the cable, in bad condition from the start, was bound to fail eventually). Some newspapers treated the entire existence of the transatlantic telegraph as a hoax or an investment scam. For the next six years, communications across the ocean would be carried much as they had been for the previous four hundred, on boats. Not until 1866 was a cable laid that held.

And were all of these lessons on the minds of Claude Shannon and his colleagues, ninety years later? Very much so: when Arthur C. Clarke paused from science fiction to write a history of communication beginning with the transatlantic cable, he dedicated it to Shannon's boss at Bell Labs, John Pierce, who "bullied" him into the project. In particular, the fiasco of the telegraph helped to crystallize three enduring lessons that would remain at the heart of communications science long after its details were forgotten, and long after the specific problem of transatlantic telegraphy had been tolerably solved.

First, communication is a war against noise. Noise is interference between telephone wires, or static that interrupts a radio transmission, or a telegraph signal corrupted by failing insulation and decaying on its way across an ocean. It is the randomness that creeps into our conversations, accidentally or deliberately, and blocks our understanding. Across short distances, or over relatively uncomplicated media—Bell calling Watson from the next room, or a landline telegraph from London to Manchester—noise could be coped with. But as distances increased and the means of sending and storing messages proliferated, the problems of noise grew with them. And the provisional solutions—whether closer to Thomson's solution of listening more closely or Whitehouse's solution of shouting louder—were ad hoc and distinct from source to source, put into practice as engineers stumbled upon them. At certain distances, or in certain channels of communication, perfect accuracy looked impossible: communication would be permanently linked to doubt. Until Claude Shannon, few people, if any, suspected that there could be a unified answer to noise.

Second, there are limits to brute force. Applying more power, amplifying messages, strengthening signals—Whitehouse's solution to the telegraph problem—remained the most intuitive answer to noise. Its failure in 1858 discredited Whitehouse but not the outlines of his methods; few others were available. Yet there were high costs to shouting. In the best case, it was still expensive and energy-hungry. In the worst case, as with the undersea cable, it could destroy the medium of communication itself.

Third, what hope there was of doing better lay in investigating the boundaries between the hard world of physics and the invisible world of messages. The object of study was the relationship between the qualities of messages—their susceptibility to noise, the density of their content, their speed, their accuracy—and the physical media that carried them. Thomson's proposed law of squares was one of the earliest links in that chain of thought. But such a law addressed only the movement of electricity, not the nature of the messages it carried. How could science speak of such a thing? It could track the speed of electrons in a wire, but the idea that the message they represented could be measured and manipulated with comparable precision would have to wait until the next century. Information was old. A science of information was just beginning to stir.

15

From Intelligence to Information

Information was something guessed at rather than spoken of, something implied in a dozen ways before it was finally tied down. Information was a presence offstage. It was there in the studies of the physiologist Hermann von Helmholtz, who, electrifying frog muscles, first timed the speed of messages in animal nerves just as Thomson was timing the speed of messages in wires. It was there in the work of physicists like Rudolf Clausius and Ludwig Boltzmann, who were pioneering ways to quantify disorder—entropy—little suspecting that information might one day be quantified in the same way. Above all, information was in the networks that descended in part from that first attempt to bridge the Atlantic. In the attack on the practical engineering problems of connecting Points A and B—what is the smallest number of wires we need to string up to handle a day's load of messages? how do we encrypt a top-secret telephone call?—the properties of information itself, in general, were gradually uncovered.

By the time of Claude Shannon's childhood, the world's communications networks were no longer passive wires acting as conduits for electricity, a kind of electron plumbing, as they were in Thomson's day. They were continent-spanning machines, arguably the most complex machines in existence. Vacuum-tube amplifiers strung along the telephone lines added power to voice signals that would have otherwise attenuated and died out on their thousand-mile journeys. A year before Shannon was born, in fact, Bell and Watson inaugurated the transcontinental phone line by reenacting their first call, this time with Bell in New York and Watson in San Francisco. By the time Shannon was a young wig-wag signaling champion, feedback systems managed the phone network's amplifiers automatically, holding the voice signals stable and silencing the "howling" or "singing" noises that plagued early phone calls, even as the seasons turned and the weather changed around the sensitive wires that carried them. Each year that Shannon placed a call, he was less likely to speak to a human operator and more likely to have his call placed by machine, by one of the automated switchboards that Bell Labs grandly called a "mechanical brain." In the process of assembling and refining these sprawling machines, Shannon's generation of scientists came to understand information in much the same way that an earlier generation of scientists came to understand heat in the process of building steam engines.

It was Shannon who made the final synthesis, who defined the concept of information and effectively solved the problem of noise. It was Shannon who was credited with gathering the threads into a new science. But he had important predecessors at Bell Labs, two engineers who had shaped his thinking since he discovered their work in Ann Arbor, who were the first to consider how information might be put on a scientific footing, and whom Shannon's landmark paper singled out as pioneers.

One was Harry Nyquist. When he was eighteen, his family left its Swedish farm and joined the wave of Scandinavian immigration to the upper Midwest; he worked construction in Sweden for four years to pay for his share of the passage. Ten years after his arrival, he had a doctorate in physics from Yale and a job as a scientist in the Bell System. A Bell lifer, Nyquist was responsible for one of the first prototype fax machines: he sketched out a proposal for "telephotography" as early as 1918. By 1924 there was a working model: a machine that scanned a photograph, represented the brightness of each chunk with its own level of electrical current, and sent those currents in pulses over the phone lines, where they were retranslated into a photographic negative on the other end, ready for the darkroom. Impressive as the display was, the market had little appetite for it, especially with its seven minutes of transmission time for a single small photo. But Nyquist's thoughts on a less glamorous technology, the telegraph, were published in the same year. Those insights would prove far more lasting.

By the 1920s, telegraphy was an old technology; it had not been at the leading edge of innovation for decades. The exciting hardware developments were in telephone networks and even, as Nyquist showed, in telephotography—applications that made use of continuous signals, while the telegraph could only speak in dot and dash. Yet the Bell System still operated a massive telegraph network, and money and careers were still riding on the same problems with which Thomson had grappled: how to send signals through that network at a maximum of speed and a minimum of noise.

Engineers already understood, Nyquist recalled, that the electrical signals carrying messages through networks—whether telegraph, -phone, or -photo—fluctuated wildly up and down. Represented on paper, the signals would look like waves: not calmly undulating sine waves, but a chaotic, wind-lashed line seemingly driven without a pattern. Yet there *was* a pattern. Even the most anarchic fluctuation could be resolved into the sum of a multitude of calm, regular waves, all crashing on top of one another at their own frequencies until they frothed into chaos. (This was the same math, in fact, that revealed tidal fluctuations to be the sum of many simple functions, and so helped make possible the first analog computers.) In this way, communications networks could carry a range, or a "band," of frequencies. And it seemed that a greater range of frequencies imposed on top of one another, a greater "bandwidth," was needed to generate the more interesting and complex waves that could carry richer information. To efficiently carry a phone conversation, the Bell network needed frequencies ranging from about 200 to 3,200 hertz, or a bandwidth of 3,000 hertz. Telegraphy required less; television would require 2,000 times more.

Nyquist showed how the bandwidth of any communications channel provided a cap on the amount of "intelligence" that could pass through it at a given speed. But this limit on intelligence meant that distinction between continuous signals (like the message on a phone line) and discrete signals (like dots and dashes or, we might add, 0's and 1's) was much less clear-cut than it seemed. A continuous signal still varied smoothly in amplitude, but you could also represent that signal as a series of samples, or discrete time-slices—and within the limit of a given bandwidth, *no one would be able to tell the difference*. Practically, that result showed Bell Labs how to send telegraph and telephone signals on the same line without interference between the two. More fundamentally, as a professor of electrical engineering wrote, it showed that "the world of technical communications is essentially discrete or 'digital.'"

Nyquist's most important contribution to the idea of information was buried in the middle of a 1924 paper read into the record of an engineers' technical conference in Philadelphia. It was only four short paragraphs under the unpromising heading "Theoretical Possibilities Using Codes with Different Numbers of Current Values." Those four paragraphs were, it turned out, a first crack at explaining the relationship between the physical properties of a channel and the speed with which it could transmit intelligence. It was a step beyond Thomson: intelligence was not electricity.

So what was it? In Nyquist's words, "by the speed of transmission of intelligence is meant the number of characters, representing different letters, figures, etc., which can be transmitted in a given length of time." This was much less clear than it might have been—but for the first time, someone was groping toward a meaningful way of treating messages scientifically. Here, then, is Nyquist's formula for the speed at which a telegraph can send intelligence:

$$W = k \log m$$

W is the speed of intelligence. m is the number of "current values" that the system can transmit. A current value is a discrete signal that a telegraph system is equipped to send: the number of current values is something like the number of possible letters in an alphabet. If the system can only communicate "on" or "off," it has two current values; if it can communicate "negative current," "off," and "positive current," it has three; and if it can communicate "strong negative," "negative," "off," "positive," and "strong positive," it has five.* Finally, *k* is the number of current values the system is able to send each second.

In other words, Nyquist showed that the speed at which a telegraph could transmit intelligence depended on two factors: the speed at which it could send signals, and the number of "letters" in its vocabulary. The more "letters" or current values that were *possible,* the fewer that would actually have to be sent over the wire. As an extreme case, imagine that there were a single ideogram that represented the entire content of this paragraph, and another single ideogram that represented the entire content of the paragraph just above; if that were the case, then we could convey the intelligence in these paragraphs to you hundreds of times faster. That was Nyquist's surprising result: the larger the number of "letters" a telegraph system could use, the faster it could send a message. Or we can look at it the other way around. The larger the number of possible current values we can choose from, the greater the density of intelligence in each signal, or in each second of communication. In the same way, our hypothetical ideogram could carry as much intelligence as all 1,262 characters in this paragraph—but only because it would have been chosen from a dictionary of millions and millions of ideograms, each somehow representing an entire paragraph of its own.†

* Even with three, five, or more current values, such a system is still *digital:* it still uses discrete steps from one value to another (as on a digital clock), rather than a continuous sweep (as on an analog clock). Digital systems are very often binary (they have only two values, as in Shannon's discussion of switching circuits), but they don't have to be.

† Of course, the impossibility of maintaining and memorizing such a dictionary points to the price of wildly accumulating symbols or current values. What an alphabetic language loses in density and efficiency, it can gain in ease of comprehension. In the same way, there is a point at which the cost of building more current values into a telegraph system outweighs the savings of faster messaging.

Nyquist's short digression on current values offered the first hint of a connection between intelligence and choice. But it remained just that. Nyquist was more interested in engineering more efficient systems than in speculating about the nature of this intelligence; and, more to the point, he was still expected to produce some measure of practical results. So, after recommending to his colleagues that they build more current values into their telegraph networks, he turned to other work. Nor, after leaving the tantalizing suggestion that all systems of communication resembled the telegraph in their digital nature, did he go on to generalize about communication itself. At the same time, his way of defining intelligence—"different letters, figures, etc."—remained distressingly vague. Behind the letters and figures there was—what, exactly?

From *Intelligence* to *information*: such a change in names tells us little about the math that underlies them. But in this case, the renaming is a useful marker. It is a border—arbitrary, in the way that very many borders are—between the adolescence and the maturity of a new science.

Reading the work of Ralph Hartley, Shannon said, was "an important influence on my life." Not simply on his research or his studies: Shannon spent much of his life working with the conceptual tools that Hartley built, and for the better part of his life, much of his public identity—"Claude Shannon, Father of Information Theory"—was bound up in having been the one who extended Hartley's ideas far beyond what Hartley, or anyone, could have imagined. Aside from George Boole, that obscure logician, no one shaped Shannon's thought more. In the 1939 letter in which Shannon first laid out the study of communications that he would complete nine years later, he used Nyquist's "intelligence." By the time the work was finished, he used Hartley's crisper term: "information." While an engineer like Shannon would not have needed the reminder, it was Hartley who made meaning's irrelevance to information clearer than ever.

After his graduation from Oxford as one of the first Rhodes scholars, Hartley was put to work on yet another effort to bridge the Atlantic. He led the Bell System team designing receivers for the first transatlantic voice call, one sent over radio waves, not wires. This time, the hindrance was not physical, but political. By the time the test was ready, in 1915, Europe was at war. The Bell engineers had to beg the French authorities for the use of the continent's highest radio antenna, which doubled as a key military asset. In the end, the Americans were allowed just minutes of precious time atop that antenna, the Eiffel Tower, but they were enough: Hartley's receivers were a success, and a human voice sent from Virginia was heard at the top of the tower.

From the beginning, Hartley's interests in communications networks were more promiscuous than Nyquist's: he was in search of a single framework that could encompass the information-transmitting power of any medium—a way of comparing telegraph to radio to television on a common scale. And Hartley's 1927 paper, which brought Nyquist's work to a higher level of abstraction, came closer to the goal than anyone yet. Suiting that abstraction, the paper Hartley presented to a scientific conference at Lake Como, in Italy, was simply called "Transmission of Information."

It was an august crowd that had assembled at the foot of the Alps for the conference. In attendance were Niels Bohr and Werner Heisenberg, two founders of quantum physics, and Enrico Fermi, who would go on to build the world's first nuclear reactor, under the bleacher seats at the University of Chicago's stadium—and Hartley was at pains to show that the study of information belonged in their company. He began by asking his audience to consider a thought experiment. Imagine a telegraph system with three current values: negative, off, and positive. Instead of allowing a trained operator to select the values with his telegraph key, we hook the key up to a random device, say, "a ball rolling into one of three pockets." We roll the ball down the ramp, send a random signal, and repeat as many times as we'd like. We've sent a message. Is it meaningful?

It depends, Hartley answered, on what we mean by meaning. If the wire was sound and the signal undistorted, we've sent a clear and readable set of symbols to our receiver—much clearer, in fact, than a human-generated message over a faulty wire. But however clearly it comes through, the message is also probably gibberish: "The reason for this is that only a limited number of the possible sequences have been assigned meanings," and a random choice of sequence is far more likely to be outside that limited range. We've arbitrarily agreed that the sequence *dot dot dot dot, dot, dot dash dot dot, dot dash dot dot, dash dash dash* carries meaning, while the sequence *dot dot dot dot, dot, dot dash dash dot, dot dash dot dot, dash dash dash* carries nonsense.* There's only meaning where there's prior agreement about our symbols. And *all* communication is like this, from waves sent over electrical wires, to the letters agreed upon to symbolize words, to the words agreed upon to symbolize things.

For Hartley, these agreements on the meaning of symbol vocabularies all depend on "psychological factors"—and those were two dirty words. Some symbols were relatively fixed (Morse code, for instance), but the meaning of many others varied with language, personality, mood, tone of voice, time of day, and so much more. There was no precision there. If, following Nyquist, the quantity of information had something to do with choice from a number of symbols, then the first requirement was getting to clarity on the number of symbols, free from the whims of psychology. A science of information would have to make sense of the messages we call gibberish, as well as the messages we call meaningful. So in a crucial passage, Hartley explained how we might begin to think about information not psychologically, but physically: "In estimating the capacity of the physical system to transmit information we should ignore the question of interpretation, make each selection perfectly arbitrary, and base our results on the possibility of the receiver's distinguishing the result of selecting any one symbol from that of selecting any other."

In this, Hartley formalized an intuition already wired into the phone company—which was, after all, in the business of transmission, not interpretation. As in the thought experiment of a telegraph

---

* Decoded from Morse code, the first sequence reads as "hello," the second as "heplo." Still, the receiver might recognize "heplo" as a typo or transmission error thanks to the *redundancy* of our language—an idea that would prove highly useful to Shannon.

controlled by a rolling ball, the only requirements are that the symbols make it through the channel, and that someone at the other end can tell them apart.

The real measure of information is not in the symbols we send—it's in the symbols we could have sent, *but did not*. To send a message is to make a selection from a pool of possible symbols, and "at each selection there are eliminated all of the other symbols which might have been chosen." To choose is to kill off alternatives. We see this most clearly, Hartley observed, in the cases in which messages happen to bear meaning. "For example, in the sentence, 'Apples are red,' the first word eliminated other kinds of fruit and all other objects in general. The second directs attention to some property or condition of apples, and the third eliminates other possible colors." This rolling process of elimination holds true for any message. The information value of a symbol depends on the number of alternatives that were killed off in its choosing. Symbols from large vocabularies bear more information than symbols from small ones. Information measures freedom of choice.

In this way, Hartley's thoughts on choice were a strong echo of Nyquist's insight into current values. But what Nyquist demonstrated for telegraphy, Hartley proved true for *any* form of communication; Nyquist's ideas turned out to be a subset of Hartley's. In the bigger picture, for those discrete messages in which symbols are sent one at a time, only three variables controlled the quantity of information: the number k of symbols sent per second, the size s of the set of possible symbols, and the length n of the message. Given these quantities, and calling the amount of information transmitted H, we have:

$$H = k \log s^n$$

If we make random choices from a set of symbols, the number of possible messages increases exponentially as the length of our message grows. For instance, in our 26-letter alphabet there are 676 possible two-letter strings (or 26^2), but 17,576 three-letter strings (or 26^3). Hartley, like Nyquist before him, found this inconvenient. A measure of information would be more workable if it increased linearly with each additional symbol, rather than exploding exponentially. In this way, a 20-letter telegram could be said to hold twice as much information as a 10-letter telegram, provided that both messages used the same alphabet. That explains what the logarithm is doing in Hartley's formula (and Nyquist's): it's converting an exponential change into a linear one. For Hartley, this was a matter of "practical engineering value."*

Engineering value was indeed what he was after, despite efforts to pin down information that sounded more like those of a philosopher or a linguist. What is the nature of communication? What happens when we send a message? Is there information in a message you can't even understand? These were powerful questions in their own right. But in all the generations of human communication, those questions were posed with urgency and rigor just *then* because the answers had suddenly grown exceptionally valuable. In the profusion of undersea cables, transcontinental radio calls, pictures sent by phone line, and moving images passing through the air, our sudden skill at communicating had outstripped our knowledge of communication itself. And whether in disaster—a fried cable—or merely an inconvenience—the flicker and blur of the first televisions—that ignorance exacted its toll.

Hartley came the nearest thus far to the essence of information. More than that, his work reflected the dawning awareness that clarity about information was already extending engineers' powers. For instance, they could chop up continuous signals, such as the human

---

* It's entirely fair to design a new measurement with human needs in mind, as long as the measurement is internally consistent. By comparison, there's no natural reason why a single degree Celsius should cover a wider range of temperature than a single degree Fahrenheit—it's just that many people find it convenient to think of water's freezing point as 0° and its boiling point as 100° and define the degrees in between accordingly. Choosing whether to think of information as a quantity that increases exponentially or linearly with message length is a matter of human convenience in the same way, which is why Shannon would describe the logarithmic scale for information as "nearer to our intuitive feelings as to the proper measure."

voice, into digital samples—and with that done, the information content of any message, continuous or discrete, could be held to a single standard. How much information, for instance, is in a picture? We can think of a picture just as we think of a telegraph. In the same way we can break a telegraph into a discrete string of dots and dashes, we can break a picture into a discrete number of squares that Hartley called "elementary areas": what were later termed picture elements, or pixels. Just as telegraph operators choose from a finite set of symbols, each elementary area is defined by a choice from a finite number of intensities. The larger the set of intensities, and the larger the number of elementary areas, the more information the picture holds. That explains why color images hold more information than images in black and white—the choice made in each pixel comes from a larger vocabulary of symbols.

Squares and intensities: the image might be the Last Supper or a dog's breakfast, but information is indifferent. In this notion that even a picture can be quantified, there's an insight into information's radically utilitarian premises, its almost Faustian exchange. But when we accept those premises, we have the first inklings of a unity behind every message.

And if some humans can achieve indifference to meaning only with great, practically ascetic effort, our machines are wired for this indifference: they have it effortlessly. So a common measure of information might allow us to express the limits of our machines and the content of our human messages in the same equations—how to shape machines and messages to a common fit. A measure for information, for example, helps us uncover the connections between the bandwidth of a medium, and the information in the message, and the time devoted to sending it. As Hartley showed, there is always a trade-off between these three quantities. To send a message faster, we can pay for more bandwidth or simplify the message. If we save on bandwidth, we pay the price in less information or a longer transmission time. This explained why, in the 1920s, sending an image over phone lines took so impractically long: phone lines lacked the bandwidth for something so complicated. Treating information, bandwidth, and time as three precise, swappable quantities could show which ideas for sending messages were "within the realm of physical possibility"—and which shouldn't even be attempted.

Last, clarity about information might lead to clarity about noise. Noise might be something more precise than the crackle of static or a series of electric pulses lost somewhere under the Atlantic; noise might be measurable, too. Hartley ventured only part of the way toward this goal, but he shed light on a specific kind of distortion he called "intersymbol interference." If the main criterion for a valid message was that the receiver tells the symbols apart, then an especially worrisome kind of inaccuracy was the type that causes symbols to blur into unreadability, as in the overlap of telegraph pulses sent by an overeager operator. With a measurement of information, we might calculate not only the time required to send any message over a given bandwidth, but the number of symbols we can send each second before they arrive too quickly to be distinguished.

This, then, was roughly where information sat when Claude Shannon picked up the thread. What began in the nineteenth century as an awareness that we might speak to one another more accurately at a distance if we could somehow quantify our messages had—almost—ripened into a new science. Each step was a step into higher abstraction. Information was the electric flow through a wire. Information was a number of characters sent by a telegraph. Information was choice among symbols. At each iteration, the concrete was falling away.

As Shannon chewed all of this over for a decade in his bachelor's apartment in the West Village or behind his closed door at Bell Labs, it seemed as if the science of information had nearly ground to a halt. Hartley himself was still on the job at Bell Labs, a scientist nearing retirement when Shannon signed on, but too far out of the mainstream for the two to collaborate effectively. The Hartley whom Shannon finally met in person seemed far removed from the Hartley who had captivated him in school. Shannon remembered him as

> very bright in some ways, but in some ways he got hung up on things. He was kind of hung up on a theory that Einstein was wrong. That Newtonian classical physics could be rescued, you

see. And he was spending all his time trying to explain all the things that relativity explained by changing the picture, just as people did . . . back in the 1920s or so, but the scientific community had finally come around to realizing that Einstein was right. All the scientific community except Hartley I guess.

So from Hartley to Shannon, said Bell Labs' John Pierce, the science of information "appears to have taken a prolonged and comfortable rest." Blame Hartley's relativity fixation, perhaps. Or blame the war—a war that unleashed tremendous applications in plane-tracking robot bombs and digital telephony, in code making and codebreaking and computing, but a war that saw few scientists with the time or incentive to step back and ask what had been learned about communication in general. Or simply blame the fact that the next and decisive step after Hartley could only be found with genius and time. We can say, from our hindsight, that if the step were obvious, it surely wouldn't have stayed untaken for twenty years. If the step were obvious, it surely wouldn't have been met with such astonishment.

"It came as a bomb," said Pierce.

16

The Bomb

The fundamental problem of communication is that of reproducing at one point either exactly or approximately a message selected at another point. Frequently the messages have *meaning*. . . . These semantic aspects of communication are irrelevant to the engineering problem."

From the start, "A Mathematical Theory of Communication" demonstrated that Shannon had digested what was most incisive from the pioneers of information science. Where Nyquist used the vague concept of "intelligence" and Hartley struggled to explain the value of discarding the psychological and semantic, Shannon took it for granted that meaning could be ignored. In the same way, he readily accepted that information measures freedom of choice: what makes messages interesting is that they are "*selected from a set* of possible messages." It would satisfy our intuitions, he agreed, if we stipulated that the amount of information on two punch cards doubled (rather than squared) the amount of information on one, or that two electronic channels could carry twice the information of one.

That was Shannon's debt. What he did next demonstrated his ambition. Every system of communication—not just the ones existing in 1948, not just the ones made by human hands, but every system conceivable—could be reduced to a radically simple essence.

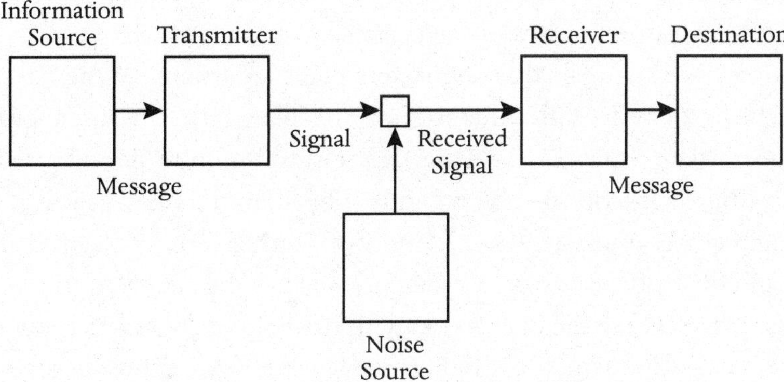

- The *information source* produces a message.
- The *transmitter* encodes the message into a form capable of being sent as a signal.
- The *channel* is the medium through which the signal passes.
- The *noise source* represents the distortions and corruptions that afflict the signal on its way to the receiver.
- The *receiver* decodes the message, reversing the action of the transmitter.
- The *destination* is the recipient of the message.

The beauty of this stripped-down model is that it applies universally. It is a story that messages cannot help but play out—human messages, messages in circuits, messages in the neurons, messages in the blood. You speak into a phone (source); the phone encodes the sound pressure of your voice into an electrical signal (transmitter); the signal passes into a wire (channel); a signal in a nearby wire interferes with it (noise); the signal is decoded back into sound (receiver); the sound reaches the ear at the other end (destination).

In one of your cells, a strand of your DNA contains the instructions to build a protein (source); the instructions are encoded in a strand of messenger RNA (transmitter); the messenger RNA carries the code to your cell's sites of protein synthesis (channel); one of the "letters" in the RNA code is randomly switched in a "point mutation" (noise); each three-"letter" code is translated into an amino acid, protein's building block (receiver); the amino acids are bound into a protein chain, and the DNA's instructions have been carried out (destination).

It is wartime. Allied headquarters plans an assault on the enemy beaches (source); staff officers turn the plan into a written order (transmitter); copies of the order are sent to the front lines, by radio or courier or carrier pigeon (channel); headquarters has deliberately scrambled the message, encrypting it to look as random as possible (a kind of artificial "noise"); one copy reaches the Allies on the front lines, who remove the encryption with the help of a key and translate it into a battle plan, but another copy is intercepted by the enemy, whose cryptanalysts crack the code for themselves (receiver); the order issued at headquarters, and intercepted by the enemy, has turned into a strategy and counterstrategy for the battle to come (destination).

Those six boxes are flexible enough to apply even to the messages the world had not yet conceived of—messages for which Shannon was, here, preparing the way. They encompass human voices as electromagnetic waves that bounce off satellites and the ceaseless digital churn of the Internet. They pertain just as well to the codes written into DNA. Although the molecule's discovery was still five years in the future, Shannon was arguably the first to conceive of our genes as information bearers, an imaginative leap that erased the border between mechanical, electronic, and biological messages.

Breaking down the act of communication into these universal steps enabled Shannon to home in on each step in isolation—to consider in turn what we do when we select our messages at the source, or how the struggle against noise can be fought and won in the channel. Imagining the *transmitter* as a distinct conceptual box proved to be especially pivotal: as we will see, the work of encoding messages for transmission turned out to hold the key to Shannon's most revolutionary result. When we remember that Shannon's mind was often at its best in the presence of outrageous analogies (as, earlier, between Boole's logic and a box of switches), we can observe how this universal structure might serve as a tool for bringing promising analogies to light.

First, though, Shannon saw that information science had still failed to pin down something crucial about information: its probabilistic nature. When Nyquist and Hartley defined it as a choice from a set of symbols, they assumed that each choice from the set would be equally probable, and would be independent of all the symbols chosen previously. It's true, Shannon countered, that *some* choices are like this. But only some. We could start, he later explained, by asking "what would be the simplest source you might have, or the simplest thing you were trying to send. And I'd think of tossing a coin." A fair coin has a 50-50 chance of landing heads or tails. This simplest choice possible—heads or tails, yes or no, 1 or 0—is the most basic message that can exist. It is the kind of message that actually conforms to Hartley's way of thinking. It would be the baseline for the true measure of information.

New sciences demand new units of measurement—as if to prove that the concepts they have been talking and talking around have at last been captured by number. The new unit of Shannon's science was to represent this basic situation of choice. Because it was a choice of 0 or 1, it was a "binary digit." In one of the only pieces of collaboration Shannon allowed on the entire project, he put it to a lunchroom table of his Bell Labs colleagues to come up with a snappier name. *Binit* and *bigit* were weighed and rejected, but the winning proposal was laid down by John Tukey, a Princeton professor working at Bell. *Bit*.

One bit is the amount of information that results from a choice between two equally likely options. So "a device with two stable positions . . . can store one bit of information." The bit-ness of such a device—a switch with two positions, a coin with two sides, a digit with two states—lies not in the outcome of the choice, but in the number of possible choices and the odds of the choosing. Two such devices would represent four total choices and would be said to store two bits. Because Shannon's measure was logarithmic (to base 2—in other words, the "reverse" of raising 2 to the power of a given number), the number of bits doubled each time the number of choices offered was squared:

Bits	Choices
1	2
2	4
4	16
8	256
16	65,536

Some choices are like this. But not all coins are fair. Not all options are equally likely. Not all messages are equally probable.

So think of the example at the opposite extreme: Think of a coin with two heads. Toss it as many times as you like—does it give you *any* information? Shannon insisted that it does not. It tells you nothing that you do not already know: it resolves no uncertainty.

What does information really measure? It measures the uncertainty we overcome. It measures our chances of learning something we haven't yet learned. Or, more specifically: when one thing carries information about another—just as a meter reading tells us about a physical quantity, or a book tells us about a life—the amount of information it carries reflects the reduction in uncertainty about the object. The messages that resolve the greatest amount of uncertainty—that are picked from the widest range of symbols with the fairest odds—are the richest in information. But where there is perfect certainty, there is no information: there is nothing to be said.

"Do you swear to tell the truth, the whole truth, and nothing but the truth?" How many times in the history of courtroom oaths has the answer been anything other than "Yes"? Because only one answer is really conceivable, the answer provides us with almost no new information—we could have guessed it beforehand. That's true of most human rituals, of all the occasions when our speech is prescribed and securely expected ("Do you take this man . . . ?"). And when we separate meaning from information, we find that some of our most meaningful utterances are also our least informative.

We might be tempted to fixate on the tiny number of instances in which the oath is denied or the bride is left at the altar. But in Shannon's terms, the amount of information at stake lies not in one particular choice, but in the probability of learning something new with any given choice. A coin heavily weighted for heads will still occasionally come up tails—but because the coin is so predictable on average, it's also information-poor.

Still, the most interesting cases lie between the two extremes of utter uncertainty and utter predictability: in the broad realm of weighted coins. Nearly every message sent and received in the real world is a weighted coin, and the amount of information at stake varies with the weighting. Here, Shannon showed the amount of information at stake in a coin flip in which the probability of a given side (call it *p*) varies from 0 percent to 50 percent to 100 percent:

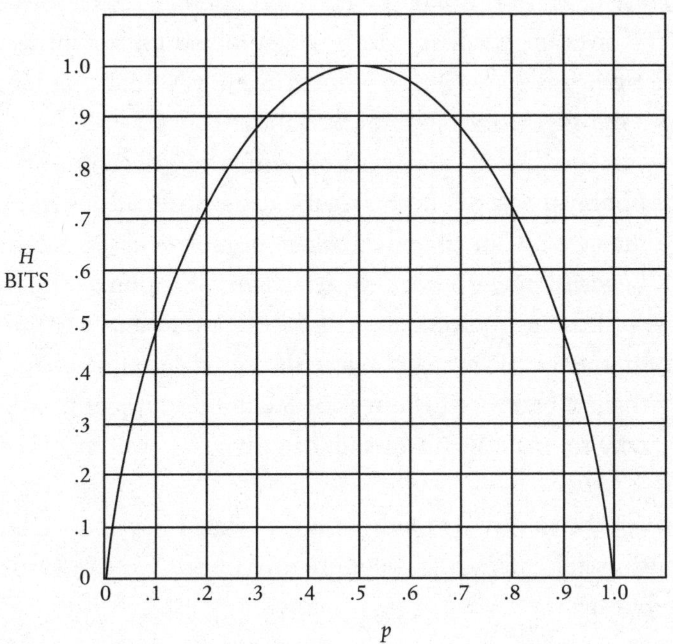

The case of 50-50 odds offers a maximum of one bit, but the amount of surprise falls off steadily as the choice grows more predictable in either direction, until we reach the perfectly predictable choice that tells us nothing. The special 50-50 case was still described by Hartley's law. But now it was clear that Hartley's theory was consumed by

Shannon's: Shannon's worked for every set of odds. In the end, the real measure of information depended on those odds:

$$H = -p \log p - q \log q$$

Here, *p* and *q* are the probabilities of the two outcomes—of either face of the coin, or of either symbol that can be sent—which together add up to 100 percent. (When more than two symbols are possible, we can insert more probabilities into the equation.) The number of bits in a message (*H*) hangs on its uncertainty: the closer the odds are to equal, the more uncertain we are at the outset, and the more the result surprises us. And as we fall away from equality, the amount of uncertainty to be resolved falls with it. So think of *H* as a measure of the coin's "average surprise." Run the numbers for a coin weighted to come up heads 70 percent of the time and you find that flipping it conveys a message worth just about .9 bits.

Now, the goal of all this was not merely to grind out the precise number of bits in every conceivable message: in situations more complicated than a coin flip, the possibilities multiply and the precise odds of each become much harder to pin down. Shannon's point was to force his colleagues to think about information in terms of probability and uncertainty. It was a break with the tradition of Nyquist and Hartley that helped to set the rest of Shannon's project in motion—though, true to form, he dismissed it as trivial: "I don't regard it as so difficult."

Difficult or not, it was new, and it revealed new possibilities for transmitting information and conquering noise. We can turn unfair odds to our favor.

For the vast bulk of messages, in fact, symbols do *not* behave like fair coins. The symbol that is sent now depends, in important and predictable ways, on the symbol that was just sent: one symbol has a "pull" on the next. Take an image: Hartley showed how to measure its information content by gauging the intensity of each "elementary area." But in images that resemble anything other than TV static, intensities are not splattered randomly across the pixels: each pixel has pull. A light pixel is more likely to appear next to a light pixel, a dark next to a dark. Or, suggested Shannon, think of the simplest case of telegraph messages. (By now it was common to appeal to the telegraph as the most basic model of discrete communication, fit for simplification and study; even as the telegraph grew obsolete, it continued to live a productive afterlife in information theory papers.) Reduce the alphabet to three basic Morse characters of dot, dash, and space. Whatever the message, a dot can be followed by a dot, dash, or space; a dash can be followed by a dot, dash, or space—but a space can only be followed by a dot or dash. A space is never supposed to be followed by another space. The choice of symbols is not perfectly free. True, a random machine in charge of a telegraph key might break the rules and ignorantly send a space after a space—but nearly all the messages that interest engineers *do* come with implicit rules, *are* something less than free, and Shannon taught engineers how to take huge advantage of this fact.

This was the hunch that Shannon had suggested to Hermann Weyl in Princeton in 1939, and which he had spent almost a decade building into theory: Information is stochastic. It is neither fully unpredictable nor fully determined. It unspools in roughly guessable ways. That's why the classic model of a stochastic process is a drunk man stumbling down the street. He doesn't walk in the respectably straight line that would allow us to predict his course perfectly. Each lurch looks like a crapshoot. But watch him for long enough and we'll see patterns start to emerge from his stumble, patterns that we could work out statistically if we cared to. Over time, we'll develop a decent estimation of the spots on the pavement on which he's most likely to end up; our estimates are even more likely to hold if we begin with some assumptions about the general walking behavior of drunks. For instance, they tend to gravitate toward lampposts.

Remarkably, as Shannon showed, this model also describes the behavior of messages and languages. Whenever we communicate, rules everywhere restrict our freedom to choose the next letter and the next pineapple.* Because these rules render certain patterns more likely and certain patterns almost impossible, languages like English come well short of complete uncertainty and maximal information: the sequence "th" has already occurred 6,431 times in this book, the sequence "tk" just this once. From the perspective of the information theorist, our languages are hugely predictable—almost boring.

To prove it, Shannon set up an ingenious, if informal, experiment in garbled text: he showed how, by playing with stochastic processes, we can construct something resembling the English language from scratch. Shannon began with complete randomness. He opened a book of random numbers, put his finger on one of the entries, and wrote down the corresponding character from a 27-symbol "alphabet" (26 letters, plus a space). He called it "zero-order approximation." Here's what happened:

> XFOML RXKHRJFFJUJ ZLPWCFWKCYJ FFJEYVKCQSGHYD QPAAMKBZAACIBZLHJQD.

There are equal odds for each character, and no character exerts a "pull" on any other. This is the printed equivalent of static. This is what our language would look like if it were perfectly uncertain and thus perfectly informative.

But we do have some certainty about English. For one, we know that some letters are likelier than others. A century before Shannon, Samuel Morse (inspired by some experimental rifling through a typesetter's box of iron characters) had built his hunches about letter frequency into his telegraph code, assigning "E" an easy single dot and "Q" a more cumbersome dash-dash-dot-dash. Morse got it roughly right: by Shannon's time, it was known that about 12 percent of English text is the letter "E," and just 1 percent the letter "Q." With a table of letter frequencies in one

* Because you're unconsciously aware of those rules, you've already recognized "pineapple" as a transmission error. Given the way the paragraph and the sentence were developing, practically the only word possible in that location was "word."

hand and his book of random numbers in the other, Shannon restacked the odds for each character. This is "first-order approximation":

> OCRO HLI RGWR NMIELWIS EU LL NBNESEBYA TH EEI ALHENHTTPA OOBTTVA NAH BRL.

More than that, though, we know that our freedom to insert any letter into a line of English text is also constrained by the character that's just come before. "K" is common after "C," but almost impossible after "T." A "Q" demands a "U." Shannon had tables of these two-letter "digram" frequencies, but rather than repeat the cumbersome process, he took a cruder tack, confident that his point was still made. To construct a text with reasonable digram frequencies, "one opens a book at random and selects a letter at random on the page. This letter is recorded. The book is then opened to another page and one reads until this letter is encountered. The succeeding letter is then recorded. Turning to another page this second letter is searched for and the succeeding letter is recorded, etc." If all goes well, the text that results reflects the odds with which one character follows another in English. This is "second-order approximation":

> ON IE ANTSOUTINYS ARE T INCTORE ST BE S DEAMY ACHIN D ILONASIVE TUCOOWE AT TEASONARE FUSO TIZIN ANDY TOBE SEACE CTISBE.

Out of nothing, a stochastic process has blindly created five English words (six, if we charitably supply an apostrophe and count ACHIN'). "Third-order approximation," using the same method to search for trigrams, brings us even closer to passable English:

> IN NO IST LAT WHEY CRATICT FROURE BIRS GROCID PONDENOME OF DEMONSTURES OF THE REPTAGIN IS REGOACTIONA OF CRE.

Not only are two- and three-letter combinations of letters more likely to occur together, but so are entire strings of letters—in other words, words. Here is "first-order word approximation," using the frequencies of whole words:

> REPRESENTING AND SPEEDILY IS AN GOOD APT OR COME CAN DIFFERENT NATURAL HERE HE THE A IN CAME THE TO OF TO EXPERT GRAY COME TO FURNISHES THE LINE MESSAGE HAD BE THESE.

Even further, our choice of the next word is strongly governed by the word that has just gone before. Finally, then, Shannon turned to "second-order word approximation," choosing a random word, flipping forward in his book until he found another instance, and then recording the word that appeared next:

> THE HEAD AND IN FRONTAL ATTACK ON AN ENGLISH WRITER THAT THE CHARACTER OF THIS POINT IS THEREFORE ANOTHER METHOD FOR THE LETTERS THAT THE TIME OF WHO EVER TOLD THE PROBLEM FOR AN UNEXPECTED.

"The particular sequence of ten words 'attack on an English writer that the character of this' is not at all unreasonable," Shannon observed with pride.*

---

* In an unpublished spoof written a year later, Shannon imagined the damage his methods would do if they fell into the wrong hands. It seems that an evil Nazi scientist, Dr. Hagen Krankheit, had escaped Germany with a prototype of his *Müllabfuhrwortmaschine,* a fearsome weapon of war "anticipated in the work . . . of Dr. Claude Shannon." Krankheit's machine used the principles of randomized text to totally automate the propaganda industry. By randomly stitching together agitprop phrases in a way that approximated human language, the *Müllabfuhrwortmaschine* could produce an endless flood of demoralizing statements. On one trial run, it spat out "Subversive elements were revealed to be related by marriage to a well-known columnist,""Capitalist warmonger is a weak link in atomic

From gibberish to reasonable, the passages grew closer and closer to passable text. They were not written, but generated: the only human intervention came in manipulating the rules. How, Shannon asked, do we get to English? We do it by making our rules more restrictive. We do it by making ourselves more predictable. We do it by becoming *less* informative. And these stochastic processes are just a model of the unthinking choices we make whenever we speak a sentence, whenever we send any message at all.

It turns out that some of the most childish questions about the world—"Why don't apples fall upwards?"—are also the most scientifically productive. If there is a pantheon of such absurd and revealing questions, it ought to include a space for Shannon's: "Why doesn't anyone say XFOML RXKHRJFFJUJ?" Investigating that question made clear that our "freedom of speech" is mostly an illusion: it comes from an impoverished understanding of freedom. Freer communicators than us—free, of course, in the sense of uncertainty and information—*would* say XFOML RXKHRJFFJUJ. But in reality, the vast bulk of possible messages have already been eliminated for us before we utter a word or write a line. Or, to alter just slightly one of the fortuitous sequences that emerged by chance on Shannon's notepad: THE LINE MESSAGE HAD [TO] BE THESE.

Still, who cares about letter frequencies?

For one, cryptanalysts do—and Shannon was one of the best. He was familiar with the charts of letter and digram and trigram frequencies because they were the codebreaker's essential tool kit. In nearly any code, certain symbols will predominate, and these symbols are likely to stand for the most common characters. Recall how, in

---

security," and "Atomic scientist is said to be associated with certain religious and racial groups." Remarkably, these machine-generated phrases were indistinguishable from human propaganda—and now it was feared that the machine had fallen into the hands of the communists.

Shannon's favorite childhood story, "The Gold-Bug," the eccentric Mr. Legrand uncovered a buried treasure by cracking this seemingly impenetrable block of code:

> 53‡‡†305))6*;4826)4‡.)4‡);806*;48†8'60))85;]8*:‡*8†83
> (88)5*†;46(;88*96*?;8)*‡(;485);5*†2:*‡(;4956*2(5*-4)8'8*; 40
> 69285);)6†8)4‡‡;1(‡9;48081;8:8‡1;48†85;4)485†528806*81
> (‡9;48;(88;4(‡?34;48)4‡;161;:188;‡?;

He began, as all good codebreakers did, by counting frequencies. The symbol "8" occurred more than any other, 33 times. This small fact was the crack that brought the entire structure down. Here, in words that captivated Shannon as a boy, is how Mr. Legrand explained it:

> Now, in English, the letter which most frequently occurs is e . . . An individual sentence of any length is rarely seen, in which it is not the prevailing character. . . .
>
> As our predominant character is 8, we will commence by assuming it as the e of the natural alphabet. . . .
>
> Now, of all words in the language, "the" is the most usual; let us see, therefore, whether they are not repetitions of any three characters in the same order of collocation, the last of them being 8. If we discover repetitions of such letters, so arranged, they will most probably represent the word "the." On inspection, we find no less than seven such arrangements, the characters being ;48. We may, therefore, assume that the semicolon represents t, that 4 represents h, and that 8 represents e—the last being now well confirmed. Thus a great step has been taken.

Being the work of a semiliterate pirate, the code was easy enough to break. More sophisticated ciphers would employ any number of stratagems to foil frequency counts: switching code alphabets partway through a message, eliminating double vowels and double consonants, simply doing without the letter "e." The codes that Shannon tested for Roosevelt and that Turing cracked for Churchill were more convoluted still. But in the end, codebreaking remained possible, and remains so, because every message runs up against a basic reality of human communication. It always involves redundancy; to communicate is to make oneself predictable.

This was the age-old codebreaker's intuition that Shannon formalized in his work on information theory: codebreaking works because our messages are less, much less, than fully uncertain. To be sure, it was not that Shannon's work in cryptography drove his breakthrough in information theory: he began thinking about information years before he began thinking about codes in any formal sense—before, in fact, he knew that he'd be spending several years as a cryptographer in the service of the American government. At the same time, his work on information and his work on codes grew from a single source: his interest in the unexamined statistical nature of messages, and his intuition that a mastery of this nature might extend our powers of communication. He would explain later, "I wrote [the information theory paper], which in a sense sort of justified some of the time I'd been putting into [cryptography], at least in my mind. . . . But there was this close connection. I mean they are very similar things. . . . Information, at one time trying to conceal it, and at the other time trying to transmit it."

In Shannon's terms, the feature of messages that makes code-cracking possible is redundancy. A historian of cryptography, David Kahn, explained it like this: "Roughly, redundancy means that more symbols are transmitted in a message than are actually needed to bear the information." Information resolves our uncertainty; redundancy is every part of a message that tells us nothing new. Whenever we can guess what comes next, we're in the presence of redundancy. Letters can be redundant: because Q is followed almost automatically by U, the U tells us almost nothing in its own right. We can usually discard it, and many more letters besides. As Shannon put it, "MST PPL HV LTTL DFFCLTY N RDNG THS SNTNC."

Words can be redundant: "the" is almost always a grammatical formality, and it can usually be erased with little cost to our understanding. Poe's cryptographic pirate would have been wise to slash the redundancy of his message by cutting every instance of "the," or ";48"—it was the very opening that Mr. Legrand exploited to such effect. Entire messages can be redundant: in all of those weighted-coin cases in which our answers are all but known in advance, we can speak and speak and say nothing new. On Shannon's understanding of information, the redundant symbols are all of the ones we can do without—every letter, word, or line that we can strike with no damage to the information.

As his approximations of text grew more and more like English, then, they also grew more and more redundant. And if this redundancy grows out of the rules that check our freedom, it is also dictated by the practicalities of communicating with one another. Every human language is highly redundant. From the dispassionate perspective of the information theorist, the *majority* of what we say—whether out of convention, or grammar, or habit—could just as well go unsaid. In his theory of communication, Shannon guessed that the world's wealth of English text could be cut in half with no loss of information: "When we write English, half of what we write is determined by the structure of the language and half is chosen freely." Later on, his estimate of redundancy rose as high as 80 percent: only one in five characters actually bear information.

As it is, Shannon suggested, we're lucky that our redundancy isn't any higher. If it were, there wouldn't be any crossword puzzles. At zero redundancy, in a world in which RXKHRJFFJUJ is a word, "any sequence of letters is a reasonable text in the language and any two dimensional array of letters forms a crossword puzzle." At higher redundancies, fewer sequences are possible, and the number of potential intersections shrinks: if English were much more redundant, it would be nearly impossible to make puzzles. On the other hand, if English were a bit less redundant, Shannon speculated, we'd be filling in crossword puzzles in three dimensions.

Shannon's estimates of our language's redundancy grew, he wrote cryptically, out of "certain known results in cryptography." The hint he dropped there is a reminder that his great work on code writing, "Communication Theory of Secrecy Systems," was still classified in 1948. Other sources, though, Shannon could discuss more openly. One was Raymond Chandler.

One evening, Shannon picked up Chandler's pulpy book of detective stories, *Pickup on Noon Street,* and flipped, as he often did in those days, to a random passage. His job was to spell the text out letter by letter; the job of his assistant was to guess the next letter until she got it right. By the time they arrived at "A S-M-A-L-L O-B-L-O-N-G R-E-A-D-I-N-G L-A-M-P O-N T-H-E D" she could guess the next three letters with perfect accuracy. E-S-K.

The point was not the assistant's powers of prediction—it was that any English reader would be just as clairvoyant in the same position, reading the same sentence governed by the same silent rules. By the time the reader has reached D, she has already gotten the point. E-S-K is a formality; and if the rules of our language left us free to shut up once the point has been gotten, D would be enough. The redundancy went even further. A phrase beginning "a small oblong reading lamp on the" is almost certainly followed by one of two letters: D, or the first guess, T. In a zero-redundancy language, the assistant would have had just a 1-in-26 chance of guessing what came next, and so the next letter would have been as informative as possible. In our language, though, her odds were much closer to 1-in-2, and the letter bore far less information. And even further: the *Oxford English Dictionary* lists 228,132 words. Out of that twenty-volume trove of lexicography, two words were hugely probable after the short phrase that Shannon spelled out: "desk"; "table." Once Raymond Chandler got to "the," he had written himself into a corner. Not that he bore any fault for it: we all write and talk and sing ourselves into corners as a condition of writing and talking and singing.

Understanding redundancy, we can manipulate it deliberately, just as an earlier era's engineers learned to play tricks with steam and heat.

Of course, humans had been experimenting with redundancy in their trial-and-error way for centuries. We cut redundancy when we write shorthand, when we assign nicknames, when we invent jargon to compress a mass of meaning ("the left-hand side of the boat when you're facing the front") into a single point ("port"). We add redundancy when we say "V as in Victor" to make ourselves more clearly heard, when we circumlocute around the obvious, even when we repeat ourselves. But it was Shannon who showed the conceptual unity behind all of these actions and more.

At the foundation of our Information Age—once wires and

microchips have been stripped away, once the stream of 0's and 1's has been parted—we find Shannon's two fundamental theorems of communication. Together they speak to the two ways in which we can manipulate redundancy: subtracting it, and adding it.

To begin with, how fast can we send a message? It depends, Shannon showed, on how much redundancy we can wring out of it. The most efficient message would actually resemble a string of random text: each new symbol would be as informative as possible, and thus as surprising as possible. Not a single symbol would be wasted. Of course, the messages that we want to send one another—whether telegraphs or TV broadcasts—do "waste" symbols all the time. So the speed with which we can communicate over a given channel depends on how we encode our messages: how we package them, as compactly as possible, for shipment. Shannon's first theorem proves that there is a point of maximum compactness for every message source. We have reached the limits of communication when every symbol tells us something new. And because we now have an exact measure of information, the bit, we also know how much a message can be compressed before it reaches that point of perfect singularity. It was one of the beauties of a physical idea of information, a bit to stand among meters and grams: proof that the efficiency of our communication depends not just on the qualities of our media of talking, on the thickness of a wire or the frequency range of a radio signal, but on something measurable, pin-downable, in the message itself.

What remained, then, was the work of source coding: building reliable systems to wring the excess from our all-too-humanly redundant messages at the source, and to reconstitute them at the destination. Shannon, along with MIT engineer Robert Fano, made an important start in this direction, and in an encyclopedia article he wrote some time after his famous paper, Shannon explained how a simple redundancy-eliminating code would work. It all depends, he said, on the statistical nature of messages: on the probability with which a white pixel happens next to a white pixel in an image, or on the frequencies of letters and digrams and trigrams that made those randomly generated fragments look more and more like English. Imagine that our language has only four letters: A, B, C, and D. Imagine that this language, like every other, lazes itself into patterns over time. Over time, half of the letters turn out to be A, a quarter turn out to be B, and C and D each make up an eighth. If we wanted to send a message in this language over the airwaves in 0's and 1's, what is the best code we could use?

Perhaps we opt for the obvious solution: each letter gets the same number of bits. For a four-letter language, we'd need two bits for each letter:

A = 00
B = 01
C = 10
D = 11

But we can do better. In fact, when transmission speed is such a valuable commodity (consider everything you can't do with a dial-up modem), we have to do better. And if we bear in mind the statistics of this particular language, we can. It's just a matter of using the fewest bits on the most common letters, and using the most cumbersome strings on the rarest ones. In other words, the least "surprising" letter is encoded with the smallest number of bits. Imagine, Shannon suggested, that we tried this code instead:

A = 0
B = 10
C = 110
D = 111*

To prove that this code is more efficient, we can multiply the number of bits for each letter by the chance that each letter will occur, giving us an average of bits per letter:

* Why not use 11 for C? In that case, it would be impossible to unambiguously decode a multi-symbol message. 1110, for instance, could mean either "CB" or "DA."

$$(1/2)\bullet 1 + (1/4)\bullet 2 + (1/8)\bullet 3 + (1/8)\bullet 3 = 1.75.$$

The message sent with this second code is less redundant: rather than using 2 bits per letter, we can express an identical idea with a leaner 1.75. It turns out that 1.75 is a special number in this four-letter language—it's also the amount of information, in bits, of any letter. Here, then, we've reached the limit. For this language, it's impossible to write a more efficient code. It's as information-dense as possible: not a digit is wasted. Shannon's first theorem shows that more complex sources—audio, video, TV, Web pages—can all be efficiently compressed in similar, if far more complex, ways.

Codes of this kind—pioneered by Shannon and Fano, and then improved by Fano's student David Huffman and scores of researchers since then—are so crucial because they enormously expand the range of messages worth sending. If we could not compress our messages, a single audio file would take hours to download, streaming Web video would be impossibly slow, and hours of television would demand a bookshelf of tapes, not a small box of discs. Because we *can* compress our messages, video files can be compacted to just a twentieth of their size. All of this communication—faster, cheaper, more voluminous—rests on Shannon's realization of our predictability. All of that predictability is fat to be cut; since Shannon, our signals have traveled light.

Yet they also travel under threat. Every signal is subject to noise. Every message is liable to corruption, distortion, scrambling, and the most ambitious messages, the most complex pulses sent over the greatest distances, are the most easily distorted. Sometime soon—not in 1948, but within the lifetimes of Shannon and his Bell Labs colleagues—human communication was going to reach the limits of its ambition, unless noise could be solved.

That was the burden of Shannon's second fundamental theorem. Unlike his first, which temporarily excised noise from the equation, the second presumed a realistically noisy world and showed us, within that world, the bounds of our accuracy and speed. Understanding those bounds demanded an investigation not simply of what we want to say, but of our means of saying it: the qualities of the channel over which our message is sent, whether that channel is a telegraph line or a fiber-optic cable.

Shannon's paper was the first to define the idea of *channel capacity,* the number of bits per second that a channel can accurately handle. He proved a precise relationship between a channel's capacity and two of its other qualities: bandwidth (or the range of frequencies it could accommodate) and its ratio of signal to noise. Nyquist and Hartley had both explored the trade-offs among capacity, complexity, and speed; but it was Shannon who expressed those trade-offs in their most precise, controllable form. The groundbreaking fact about channel capacity, though, was not simply that it could be traded for or traded away. It was that there is a hard cap—a "speed limit" in bits per second—on accurate communication in any medium. Past this point, which was soon enough named the Shannon limit, our accuracy breaks down. Shannon gave every subsequent generation of engineers a mark to aim for, as well as a way of knowing when they were wasting their time in pursuit of the hopeless. In a way, he also gave them what they had been after since the days of Thomson and the transatlantic cable: an equation that brought message and medium under the same laws.

This would have been enough. But it was the next step that seemed, depending on one's perspective, miraculous or inconceivable. Below the channel's speed limit, we can make our messages as accurate as we desire—for all intents, we can make them perfectly accurate, perfectly free from noise. This was Shannon's furthest-reaching find: the one Fano called "unknown, unthinkable," until Shannon thought it.

Until Shannon, it was simply conventional wisdom that noise had to be endured. The means of mitigating noise had hardly changed, in principle, since Wildman Whitehouse fried the great undersea cable. Transmitting information, common sense said, was like transmitting power. Expensively and precariously adding more power remained the best answer—shouting through the static, as it were, brute-forcing the signal-to-noise ratio by pumping out a louder signal.

Shannon's promise of perfect accuracy was something radically new.* For engineering professor James Massey, it was this promise above all that made Shannon's theory "Copernican": Copernican in the sense that it productively stood the obvious on its head and revolutionized our understanding of the world. Just as the sun "obviously" orbited the earth, the best answer to noise "obviously" had to do with physical channels of communication, with their power and signal strength. Shannon proposed an unsettling inversion. Ignore the physical channel and accept its limits: we can overcome noise by manipulating our messages. The answer to noise is not in how loudly we speak, but in how we say what we say.

How did the faltering transatlantic telegraph operators attempt to deal with the corruption of their signal? They simply repeated themselves: "Repeat, please." "Send slower." "Right. Right." In fact, Shannon showed that the beleaguered key-tappers in Ireland and Newfoundland had essentially gotten it right, had already solved the problem without knowing it. They might have said, if only they could have read Shannon's paper, "Please add redundancy."

In a way, that was already evident enough: saying the same thing twice in a noisy room is a way of adding redundancy, on the unstated assumption that the same error is unlikely to attach itself to the same place two times in a row. For Shannon, though, there was much more. Our linguistic predictability, our congenital failure to maximize information, is actually our best protection from error. A few pages ago, recall, you read that the structure of our language denies us total freedom to choose "the next letter and the next pineapple." As soon as you reached "pineapple"—really, as soon as you got to "p"—you knew that something was wrong. You had detected (and probably corrected) an error. You did it because, even without running the numbers, you have an innate grasp of the statistical structure of English. And that intuition told you that the odds of "pineapple" making sense in that sentence and paragraph were lottery-winning low. The redundancy of

* Or more accurately, of an "arbitrarily small" rate of error: an error rate as low as we want, and want to pay for.

our language corrected the error for you. On the other hand, imagine how much harder it would be to find an error in the "XFOML" language, a language in which each letter is equally likely.*

For Shannon, then, the key was once again in the code. We must be able to write codes, he showed, in which redundancy acts as a shield: codes in which no one bit is indispensable, and thus codes in which any bit can absorb the damage of noise. Once more, we want to send a message made up of the letters A through D, but this time we are less concerned with compressing the message than with passing it safely through a noisy channel. Again, we might start by trying the laziest code:

A = 00
B = 01
C = 10
D = 11

One of the worst things that noise can do—in a burst of static, interference from the atmosphere, or physical damage to the channel—is falsify bits. Where the sender says "1," the receiver hears "0," or vice versa. So if we used this code, an error to a single bit could be fatal.

* Kahn illustrates this point with a useful thought experiment. Think of a language in which any four-letter combination, from "aaaa" to "zzzz," was fair game. There would be 456,976 such combinations, more than enough to account for every word in an English dictionary. But when *any* letter combination is valid, recognizing errors becomes far more difficult. "'Xfim,' meaning perhaps 'come,' would be changed to 'xfem,' maybe meaning 'go' and, without redundancy, no alarm bells would ring." By contrast, ordinary languages benefit not just from the redundancy of context (which made "pineapple" impossible above), but from the redundancy of letters that bear no information. The loss of one dot in Morse code turns "individual" into "endividual"—but the error is easy to detect. Most English words can suffer similar errors to several letters before the sender's intention is lost.

If just one of the bits representing C flipped, C would vanish in the channel: it would emerge as B or D, with the receiver none the wiser. It would take just two such flips to turn "DAD" to "CAB."

But we can solve the problem—just as human languages have intuitively, automatically solved the same problem—by adding bits. We could use a code like this:

A = 00000
B = 00111
C = 11100
D = 11011

Now any letter could sustain damage to any one bit and still resemble itself more than any other letter. With two errors, things get fuzzier: 00011 could be either B with one flipped bit or A with two. But it takes fully three errors to turn one letter into another. Our new code resists noise in a way our first one did not, and does it more efficiently than simple repetition. We were not forced to change a single thing about our medium of communication: no yelling across a crowded room, no hooking up spark coils to the telegraph, no beaming twice the television signal into the sky. We only had to signal smarter.

As long as we respect the speed limit of the channel, there is no limit to our accuracy, no limit to the amount of noise through which we can make ourselves heard. Yes, overcoming more errors, or representing more characters, would demand more complex codes. So would combining the advantages of codes that compress and codes that guard against error: that is, reducing a message to bits as efficiently as possible, and then adding the redundancy that protects its accuracy. Coding and decoding would still exact their cost in effort and time. But Shannon's proof stood: there is *always* an answer. The answer is digital. Here Shannon completed the reimagining that began with his thesis and his switches eleven years earlier. 1's and 0's could enact the entirety of logic. 1's and 0's stood for the fundamental nature of information, an equal choice from a set of two. And now it was evident that any message could be sent flawlessly—we could communicate anything of any complexity to anyone at any distance—provided it was translated into 1's and 0's. Logic is digital. Information is digital.

So each message is kin to all messages. "Up until that time, everyone thought that communication was involved in trying to find ways of communicating written language, spoken language, pictures, video, and all of these different things—that all of these would require different ways of communicating," said Shannon's colleague Robert Gallager. "Claude said no, you can turn all of them into binary digits. And then you can find ways of communicating the binary digits." You can code any message as a stream of bits, without having to know where it will go; you can transmit any stream of bits, efficiently and reliably, without having to know where it came from. As information theorist Dave Forney put it, "bits are the universal interface."

In time, the thoughts developed in these seventy-seven pages in the *Bell System Technical Journal* would give rise to a digital world: satellites speaking to earth in binary code, discs that could play music through smudges and scratches (because storage is just another channel, and a scratch is just another noise), the world's information distilled into a black rectangle two inches across.

In time: because while Shannon had proven that the codes must be there, neither he nor anyone else had shown what they must be. Once the audacity of his work had worn off—he had, after all, founded a new field and solved most of its problems at one stroke—one consequential shortfall would dominate the conversation on Claude Shannon and Claude Shannon's theory. How long would it take to find the codes? Once found, would they even make everyday practical sense, or would it simply be cheaper to continue muddling through? Could this strange work, full of imaginary languages, messages without meaning, random text, and a philosophy that claimed to encompass and explain every signal that could possibly be sent, ever be more than an elegant piece of theorizing? In words with which any engineer could have sympathized: would it *work*?

Yet, from the other direction and in a far different spirit, there came another set of questions. They're best overheard in a conversation between Shannon and Von Neumann at Princeton, said to have taken place in 1940, when Shannon was first piecing his theory together in the midst of his failing marriage. Shannon approached the great man with his idea of information-as-resolved-uncertainty—which would come to stand at the heart of his work—and with an unassuming question. What should he call this thing? Von Neumann answered at once: say that information reduces "entropy." For one, it was a good, solid physics word. "And more importantly," he went on, "no one knows what entropy really is, so in a debate you will always have the advantage."

Almost certainly, this conversation never happened. But great science tends to generate its own lore, and the story is almost coeval with Shannon's paper. It was retold in seminars and lectures and books, and Shannon himself had to brush it away at conferences and in interviews with his usual evasive laugh. The story was retold for so long—as we are retelling it here—just because the link between information and entropy was so suggestive.* From one direction came the demand that Shannon's paper *work;* from the other, the suspicion that it hinted at truths more profound than the author himself was willing to admit.

No one knows what entropy really is. It was an overstatement; but entropy has, at least, been a multitude of things in its conceptual

* The link between information and entropy was made explicit in Shannon's paper. But a connection between information and physics was first suggested, as early as 1929, by the Hungarian physicist Leo Szilard. Briefly, Szilard resolved an old puzzle in the physics of heat: the Second Law of Thermodynamics says that entropy is constantly increasing, but what if we imagined a microscopic and intelligent being, which James Clerk Maxwell had dubbed a "demon," that tried to decrease entropy by sorting hot molecules from cold? Would that contradict the Second Law? Szilard showed that it would not: the very act of determining which molecules were which would cost enough energy to offset any savings that Maxwell's Demon proposed to achieve. In other words—*learning information* about particles costs energy. Shannon, however, had not read Szilard's work when he wrote his 1948 paper.

life—nearly as many things as information itself—some scientifically sound, and some otherwise. It has been the inability of a steam engine to do work; it has been the dissipation of heat and energy, the unalterable tendency of every part of a closed system to lapse toward lukewarm muck; it has been, more roughly but also more resonantly, the trend toward disorder, chaos. It is the always incipient mess against which we are pitted as a condition of living. James Gleick put this succintly: "Organisms organize." He went on:

> We sort the mail, build sand castles, solve jigsaw puzzles, separate wheat from chaff, rearrange chess pieces, collect stamps, alphabetize books, create symmetry, compose sonnets and sonatas, and put our rooms in order. . . . We propagate structure (not just we humans but we who are alive). We disturb the tendency toward equilibrium. It would be absurd to attempt a thermodynamic accounting for such processes, but it is not absurd to say that we are reducing entropy, piece by piece. Bit by bit.

In pursuing all of this order, we render our world *less* informative, because we reduce the amount of uncertainty available to be resolved. The predictability of our communication is, in this light, the image of a greater predictability. We are, all of us, predictability machines. We think of ourselves as incessant makers and consumers of information. But in the sense of Shannon's entropy, the opposite is true: we are sucking information out of the world.

Yet we are failing at it. Heat dissipates; disorder, in the very long run, increases; entropy, the physicists tell us, runs on an eternally upward slope. In the state of maximal entropy, all pockets of predictability would have long since failed: each particle a surprise. And the whole would read, were there then eyes to read it, as the most informative of messages.

The unsettled question: whether information-as-entropy was a misplaced and fruitless analogy, or whether it was a more or less resonant language in which to talk about the world—or whether, in fact, information itself was fundamental in a way that even a physicist could appreciate. When particles jump from state to state, is their resemblance to switches, to logic circuits, to 0's and 1's, something more than a trick of our eyes? Or put it this way: Was the quality of information something we imposed on the world, just a by-product of our messages and machines—or was it something we *found out* about the world, something that had been there all along?

These were only some of the insistent questions that trailed after Shannon's theory. Shannon himself—even while courting them with the use of such a tantalizing term, or metaphor, as "entropy"—almost always dismissed these puzzles. His was a theory of messages and transmission and communication and codes. It was enough. "You know where my interests are."

But in his insistence on this point, he ran up against a human habit much older than him: our tendency to reimagine the universe in the image of our tools. We made clocks, and found the world to be clockwork; steam engines, and found the world to be a machine processing heat; information networks—switching circuits and data transmission and half a million miles of submarine cable connecting the continents—and found the world in their image, too.

17

Building a Bandwagon

He would live to see "information" turn from the name of a theory to the name of an era. "The Magna Carta of the Information Age," *Scientific American* would call Shannon's 1948 paper decades later. "Without Claude's work, the internet as we know it could not have been created," ran a typical piece of praise. And on and on: "A major contribution to civilization." "A universal clue to solving problems in different fields of science." "I reread it every year, with undiminished wonder. I'm sure I get an IQ boost every time." "I know of no greater work of genius in the annals of technological thought."

But in 1948, the bulk of the honors were years away. At the time, the magnitude of information theory was intelligible only to a small clutch of communications engineers and mathematicians and only available in a technical journal—Bell Labs' *Bell System Technical Journal*. So it says something about the power and persuasiveness of Shannon's ideas that "A Mathematical Theory of Communication" rapidly received attention well outside the confines of the Labs and even the field of engineering, and would, in less than a decade, turn into a kind of international phenomenon—one that Shannon himself would, ironically and futilely, try to rein in.

In the months following publication of Shannon's paper, word of a breakthrough propagated through the community of communications engineers. "While, of course, Shannon was not working in a vacuum in the 1940s, his results were so breathtakingly original that even the communication specialists of the day were at a loss to understand their significance," writes information theorist R. J. McEliece. Yet it was clear, even then, that these results would reshape the field. Shannon's paper quickly became the jumping-off point for several others, which, in academic terms, is the equivalent of a round of applause. By November, only a month after the second installment of Shannon's work, two derivative papers appeared, exploring the advantages of pulse code modulation through the prism of his earlier ideas. Five other significant papers tied directly to Shannon's work came out soon thereafter.

Thus, beginning with the small but dedicated readership of the *Bell System Technical Journal*, news of information theory rippled through the mathematical and engineering worlds. It piqued the interest of one reader in particular, who would become Shannon's most important popularizer: Warren Weaver, the director of the Division of Natural Sciences at the Rockefeller Foundation, one of the principal funders of science and mathematics research in the country.

Weaver had entered Shannon's life earlier, when, with the support of Thornton Fry and Vannevar Bush, he awarded Shannon a contract to work on fire control during the war. Now he would play an even more pivotal role in Shannon's career, as the catalyst behind the book-length publication of "A Mathematical Theory of Communication"—a book that would do for the theory what a technical journal article could not.

The two had met in the fall of 1948 and discussed the theory. Weaver, perhaps through an excess of enthusiasm, foresaw a world in which information theory could help computers fight the Cold War and enable instantaneous rendering of Soviet documents into English. Inspired, he praised Shannon's work with exuberance to the head of the Rockefeller Foundation, Chester Barnard. In early 1949, Weaver sent Barnard his own layman's translation of "A Mathematical Theory of Communication."

"Weaver became the expositor of Shannon almost by accident," a recent history observed. By accident, indeed: Weaver's memo might have remained another forgotten interdepartmental missive, or an unread article in a journal, had it not been for the intervention of two men: Louis Ridenour, dean of graduate studies at the University of Illinois, and Wilbur Schramm, head of the university's Institute for Communications Research.

Ridenour had spent the early part of the twentieth century at the rich intersection of physics and geopolitics. During World War II, he worked at the renowned MIT Radiation Laboratory, commonly known as the Rad Lab. The Rad Lab began with outsize ambitions, as an effort to perfect mass-produced radar technology to defeat the German Luftwaffe's bombing runs against the British. It also had mysterious origins. Funded by Alfred Lee Loomis, the intensely private millionaire financier, attorney, and self-taught physicist, the lab was initially bankrolled entirely by Loomis himself. It created most of the radar systems used to identify German U-boats—and its network of scientists and technicians became much of the nucleus of the Manhattan Project. As Lee DuBridge, the lab's director, would later quip, "Radar won the war; the atom bomb ended it." This was the world of fighting man's physics.

Weaver met Ridenour on official business in Champaign-Urbana, Illinois, while exploring whether the Rockefeller Foundation should fund a biological sciences program at the university. He shared a copy of his rendition of Shannon's paper with Ridenour. He, in turn, passed the draft to Schramm, another one of the University of Illinois's brightest stars, whose Institute for Communications Research was beginning to lay the foundation of communications as a formal field of study. By some accounts the first communications scholar, Schramm established the now world-famous Iowa Writers' Workshop, home to authors from Robert Penn Warren to Marilynne Robinson.

Communications was, in a way, an ironic choice of field for Schramm. A botched childhood tonsillectomy had left him with a severe stutter, which so embarrassed him that, honored as the valedictorian of his high school class, he opted to play the flute instead of giving a speech. Speech difficulties notwithstanding, he graduated summa cum laude and Phi Beta Kappa from Marietta College, paused at Harvard for a master's, and completed a PhD in American literature at the University of Iowa, while undergoing treatment at a famous stammering clinic in Iowa City.

Schramm's many academic responsibilities included overseeing the University of Illinois Press—and, encouraged by Ridenour, he saw the opportunity in a version of the "Mathematical Theory" for the general public. His motives, and Ridenour's as well, were those of practiced institution builders. Schramm's institute sought credibility in any form for the emerging field of "communications studies." Ridenour knew that the University of Illinois was debating the purchase of a computer. A volume featuring Warren Weaver and Claude Shannon published by the university would be the perfect complement to the "press's new series of lectures by computer builders—setting up both Ridenour and Schramm to accomplish their respective projects."

Whatever the motives, the book became a reality. By the admittedly modest standards of university presses, the volume was a roaring success. Debuting in 1949, a year after the theory was made public, *The Mathematical Theory of Communication* sold 6,000 copies in its first four years of printing; by 1990, it had sold more than 51,000 copies, putting it among the bestselling academic books published by a university press.

The book was ultimately one part Shannon, two parts Weaver: Part I featured Shannon's original 1948 work; Parts II and III, by Weaver, attempted to explain the theory in as close to layman's terms as possible. The book's organization created the unintended impression of Weaver as a key contributor to the theory's development. Commentators and observers would, for decades, refer to the "Shannon and Weaver" theory of information or even go so far as to call Weaver a cofounder of the theory. Weaver never indulged the inaccuracies. He was quick to correct the record, telling Ridenour, "No one could recognize more keenly than I do that my own contribution is infinitesimal as compared with Shannon's."

In fact, the only issue Weaver had with the text was a concern that it might have exaggerated his role in information theory's development. Because his section was really an introduction to Shannon's work, it should have come first:

> There could very easily have been a short statement (perhaps most appropriately by myself?) apologising for coming first with my very modest bit, explaining why that is sensible, and expressing the great hope that all will be thus led on to study the really serious and important part of the book.

While Weaver fretted over appearing to claim false credit, Schramm and Ridenour celebrated. The volume's publication accomplished everything they could have hoped for. By 1952, the University of Illinois had succeeded in acquiring a digital computer, and simultaneously, it was awarded a large federal contract for the study of "communication theory."

The publication of *The Mathematical Theory of Communication* stands as one of the defining moments in the history of information theory, and not only on account of its commercial success. Even the title sent an important message: in the span of a year, Shannon's original "*A* Mathematical Theory of Communication" had become the definitive "*The* Mathematical Theory of Communication." As electrical engineer and information theorist Robert Gallager pointed out, the subtle change in the article's context, from one of several articles in a technical journal to centerpiece of a book, was a mark of supremacy. It stood for the scientific community's growing recognition that Shannon's theory stood alone.

18

Mathematical Intentions, Honorable and Otherwise

It's one of the curses of scientific discoveries that they are greeted, remarkably often, with misunderstanding or outright dismissal. Charles Darwin's geology teacher, the famed Adam Sedgwick, wrote to his student after the publication of *On the Origin of Species,* "I have read your book with more pain than pleasure. Parts of it I admired greatly, parts I laughed at till my sides were almost sore; other parts I read with absolute sorrow, because I think them utterly false and grievously mischievous." Sylvia Nasar, writing about John Nash's Nobel Prize–winning work on game theory, remarked that his ideas "seemed initially too simple to be truly interesting, too narrow to be widely applicable, and, later on, so obvious that its discovery by someone was deemed all but inevitable." Scientific revolutions are rarely unopposed.

Shannon's work, too, was coldly received in some quarters. The first major criticism, and the one with the most edge to it, came from the mathematician Joseph L. Doob. Brought from the Midwest to New York at the age of three, he was marked out as a bright student early and enrolled in New York's Ethical Culture Fieldston School. The school was unique in New York society at the time: its founder's belief that the poor deserved the highest-caliber education was radical, but the school's academic reputation also drew the well-heeled. Over the twentieth century, it would produce such alumni as Marvin Minsky, a pioneer of artificial intelligence and future colleague of Shannon's, and J. Robert Oppenheimer, the father of the atomic bomb.

After excelling there, Doob departed for Harvard, where, it is said, he grew so frustrated with the plodding pace of math instruction that he took sophomore and junior calculus simultaneously—and aced both. Unlike many of his fellow students, Doob never harbored any doubts about his future as a mathematician.

His confidence showed in the scale of the work he attempted: his 1953 book on probability theory, an 800-page tome, was greeted as the most influential work on the topic since the nineteenth century. His assurance displayed itself in another way too: Doob was a fierce critic of anything he regarded as flabby thinking. Doob was open about the fact that he was, perhaps too frequently, looking for trouble. Asked why he became interested in mathematics in the first place, he answered:

> I have always wanted to understand what I was doing, and why I was doing it, and I have often been a pest because I have objected when what I heard or read was not to be taken literally. The boy who noticed that the emperor wasn't dressed and objected loudly has always been my model. Mathematics seemed to match my psychology, a mistake reflecting the fact that somehow I did not take into account that mathematics is created by humans.

His sharp words, friends recalled, often came mixed with humor. Once, he and a colleague, Robert Kaufman, fell into a heated argument about whether students ought to be required to read classic literature. "Robert was all for it, and Joe was doing everything to provoke him. In disgust, Robert said: 'Oh my God!' and Joe calmly replied, 'Please don't exaggerate, just call me Professor.'"

Above all, Doob professed loyalty to the "austere and often abstruse" world of pure mathematics. If applied mathematics concerns itself with concrete questions, pure mathematics exists for its own sake. Its cardinal questions are not "How do we encrypt a telephone conversation?" but rather "Are there infinitely many twin primes?" or "Does every true mathematical statement have a proof?" The divorce between the two schools has ancient origins. Historian Carl Boyer traces it to Plato, who regarded mere computation as suitable for a merchant or a general, who "must learn the art of numbers or he will not know how to array his troops." But the philosopher must study higher mathematics, "because he has to arise out of the sea of change and lay hold of true being." Euclid, the father of geometry, was a touch snobbier: "There is a tale told of him that when one of his students asked of what use was the study of geometry, Euclid asked his slave to give the student threepence, 'since he must make gain of what he learns.'"

Closer to our times, the twentieth-century mathematician G. H. Hardy would write what became the ur-text of pure math. *A Mathematician's Apology* is a "manifesto for mathematics itself," which pointedly borrowed its title from Socrates's argument in the face of capital charges. For Hardy, mathematical elegance was an end in itself. "Beauty is the first test," he insisted. "There is no permanent place in the world for ugly mathematics." A mathematician, then, is not a mere solver of practical problems. He, "like a painter or a poet, is a maker of patterns. If his patterns are more permanent than theirs, it is because they are made with ideas." By contrast, run-of-the-mill applied mathematics was "dull," "ugly," "trivial," and "elementary."

It was the pure mathematicians who looked down on Von Neumann's work on game theory, calling it, among other things, "just the latest fad" and "déclassé." The same group would level a similar judgment against John Nash—just as Doob would against Claude Shannon.

As America's leading probability theorist, Doob was well positioned to review Shannon's work. His critique appeared in the pages of *Mathematical Review* in 1949. After briefly summarizing the contents of Shannon's paper, he dismissed them with a sentence that would irritate Shannon's supporters for years: "The discussion is suggestive throughout, rather than mathematical, and it is not always clear that the author's mathematical intentions are honorable." By the genteel standards of an academic review, this was lacerating, the equivalent of pistols at dawn.

Nearly forty years later, interviewer Anthony Liversidge raised the issue of the Doob critique with Shannon:

LIVERSIDGE: When *The Mathematical Theory of Communication* was published, there was an indignant review by a certain mathematician, accusing you of mathematical dishonesty because your results weren't proved, he said, with mathematical rigor. Did you think that plain silly, or did you think, "Well, maybe I should work hard to meet his criticisms?"

SHANNON: I didn't like his review. He hadn't read the paper carefully. You can write mathematics line by line with each tiny inference indicated, or you can assume that the reader understands what you are talking about. I was confident I was correct, not only in an intuitive way but in a rigorous way. I knew exactly what I was doing, and it all came out exactly right.

Shannon rarely, if ever, felt the need to defend himself; the Doob critique, in other words, had clearly gotten to him. What's more, Shannon was fully aware of his having vaulted over some of the intervening mathematics in the interest of practicality. Importantly, he noted in the middle of "A Mathematical Theory of Communication" that "the occasional liberties taken with limiting processes in the present analysis can be justified in all cases of practical interest." This made sense: his primary readers were communications engineers; practical intentions mattered as much as, if not more than, purely mathematical ones. For Doob to critique the precision of his math felt, to many of Shannon's acolytes, a bit like examining the *Mona Lisa* and finding fault with the frame.

Ironically, Doob's claim that the paper was not mathematical enough ran up against the opposite complaint from engineers. As the mathematician Solomon Golomb put it, "When Shannon's paper appeared, some communications engineers found it to be too mathematical (there are twenty-three theorems!) and too theoretical." The problem, in hindsight, might not have been Doob's misunderstanding of the mathematics at work; rather, he didn't understand that

Shannon's math was a means to an end. "In reality," said Golomb, "Shannon had almost unfailing instinct for what was actually *true* and gave outlines of proofs that other mathematicians . . . would make fully rigorous." In the words of one of Shannon's later collaborators, "Distinguished and accomplished as Doob was, the gaps in Shannon's paper which seemed large to Doob seemed like small and obvious steps to Shannon. Doob might not realize this for, how often if ever, would he have encountered a mind like Shannon's?"

The information theorist Sergio Verdú offered a similar assessment of Shannon's paper: "It turned out that everything he claimed essentially was true. The paper was weak in what we now call 'converses' . . . but in fact, that adds to his genius rather than detracting from it, because he really knew what he was doing." In a sense, leaving the dots for others to connect was a calculated gamble on Shannon's part: had he gone through that painstaking work himself, the paper would have been much longer and appeared much later, both factors that would have likely diminished its reception. By the end of the 1950s, other engineers and mathematicians in both the United States and the Soviet Union had followed Shannon's lead—and translated Shannon's creative and rigorous explanations into both the language of pure mathematicians and the language of engineers.

Criticism like Doob's may have stung—but there was a measure of respect in the fact that a mathematician of Doob's stature had read Shannon's work at all. Doob and Shannon also settled their differences in 1963. Shannon was invited by the American Mathematical Society to deliver the prestigious Josiah Willard Gibbs Lecture, a signal honor within the field. The person who introduced him on that night—and, as the society's president, surely had a hand in the invitation—was none other than Joseph L. Doob.

19

Wiener

He was, according to one writer, "the American John von Neumann"—and the exaggeration was almost excusable.

Born in Columbia, Missouri, Norbert Wiener was shaped by a father single-mindedly focused on molding his young son into a genius. Leo Wiener used an extraordinary personal library—and an extraordinary will—to homeschool young Norbert until the age of nine. "I had full liberty to roam in what was the very catholic and miscellaneous library of my father," Wiener wrote. "At one period or other the scientific interests of my father had covered most of the imaginable subjects of study."

But Leo's training was also unsparing, even cruel, and his son was denied a normal childhood. In a passage from *Ex-Prodigy: My Childhood and Youth,* his memoir, Wiener recalls his father's instruction:

> He would begin the discussion in an easy, conversational tone. This lasted exactly until I made the first mathematical mistake. Then the gentle and loving father was replaced by the avenger of the blood. . . . Father was raging, I was weeping, and my mother did her best to defend me, although hers was a losing battle.

At one point, a doctor ordered young Norbert to stop reading; his eyesight could afford no additional strain. Leo decided that what his son could not read, he could memorize. Even a doctor's concern couldn't stop Wiener's determined father: Leo would lecture endlessly, and young Norbert would be expected to capture every word and thought.

In a purely professional sense, the intense regimen of schooling paid dividends. By age eleven, Wiener had already finished high school. Three years after that, at the age of fourteen, he graduated Tufts with a degree in mathematics. From there it was on to Harvard to study zoology, Cornell to study philosophy, and finally, a return to Harvard, where, by seventeen, he earned a PhD in mathematics with a specialty in logic. The climb into the elite echelons of mathematics—and the kind of life his father might have wished for his son—had begun.

But the scars of such a childhood were obvious for all to see. He had been a child coming of age around people many years his senior. And as often happens, he was ridiculed cruelly and mercilessly by the older children; the result was an intense awkwardness that followed him his entire life. It didn't help that, in appearance, Wiener was easy to ridicule. Bearded, bespectacled, nearsighted, with red-veined skin and a ducklike walk, there was hardly a stereotype of the addle-pated academic that Wiener did not fulfill. "From every angle of vision there was something idiosyncratic about Norbert Wiener," mused Paul Samuelson. Hans Freudenthal remembered,

> In appearance and behaviour, Norbert Wiener was a baroque figure, short, rotund, and myopic, combining these and many qualities in extreme degree. His conversation was a curious mixture of pomposity and wantonness. He was a poor listener. . . . He spoke many languages but was not easy to understand in any of them. He was a famously bad lecturer.

The anecdotes about him fill the pages of other mathematicians' memoirs, and nearly all are the kind that were first shared behind Wiener's back. As one story had it, Wiener arrived at what he thought was his home and fumbled with his keys, finding that they would not fit in the lock. He turned to the children playing in the street and asked, "Can you show me where the Wieners live?" A little girl replied, "Follow me, Daddy. Mommy sent me here to point the way to our new house."

His contributions to mathematics were as broad as they were deep: quantum mechanics, Brownian motion, cybernetics, stochastic processes, harmonic analysis—there was hardly a corner of the mathematical universe that his intellect left untouched. By 1948, he had a CV packed with glittering awards and honors. Wiener's list of collaborators and contacts was similarly striking: Vannevar Bush, G. H. Hardy, Bertrand Russell, Paul Lévy, Kurt Gödel . . . and Claude Shannon.

At MIT, Shannon had taken Wiener's class in Fourier analysis. A half century later, reflecting on his time in graduate school, Shannon would remember Wiener as "an idol of mine when I was a young student." Shannon seems not have made a similar impression on Wiener, who wrote in his 1956 memoir, "Shannon and I had relatively little contact during his stay here [at MIT] as a student." He added, though, that "since then, the two of us have developed along parallel if different directions, and our scientific relations have greatly broadened and deepened."

Wiener was twenty-two years older than Shannon, so it reveals something about the advanced degree of Shannon's thinking and the importance of his work that, as early as 1945, Wiener was nervous about which of them would win the race for credit for information theory. Their contest began in earnest in 1946.

As the story goes, the manuscript that formed the outlines of Wiener's contributions to information theory was nearly lost to humanity. Wiener had entrusted the manuscript to Walter Pitts, a graduate student, who had checked it as baggage for a trip from New York's Grand Central Terminal to Boston. Pitts forgot to retrieve the baggage. Realizing his mistake, he asked two friends to pick up the bag. They either ignored or forgot the request. Only five months later was the manuscript finally tracked down; it had been labeled "unclaimed property" and cast aside in a coatroom.

Wiener was, understandably, blind with rage. "Under these

circumstances please consider me as completely dissociated from your future career," he wrote to Pitts. He complained to one administrator of the "total irresponsibleness of the boys" and to another faculty member that the missing parcel meant that he had "lost priority on some important work." "One of my competitors, Shannon of the Bell Telephone Company, is coming out with a paper before mine," he fumed. Wiener wasn't being needlessly paranoid: Shannon had, by that point, previewed his still-unpublished work at 1947 conferences at Harvard and Columbia. In April 1947, Wiener and Shannon shared the same stage, and both had the opportunity to present early versions of their thoughts. Wiener, in a moment of excessive self-regard, would write to a colleague, "The Bell people are fully accepting my thesis concerning statistics and communications engineering."

Wiener's contribution was contained within the wide-ranging book *Cybernetics*, which had its debut in the same year as Shannon's two-part paper. If Shannon's 1948 work was, at least initially, relatively unknown to the wider public, Wiener's notion of cybernetics—a word he derived from the Greek for "steersman" to encompass "the entire field of control and communications theory, whether in the machine or in the animal"—aroused intense public interest from the moment it was published. A bestseller, the book managed to find its way into the hands of nontechnical readers. The praise was fulsome, the kind of acclaim that most authors work a lifetime to achieve. In the *New York Times*, the physicist John R. Platt ranked *Cybernetics* as one of those books that "might be comparable in ultimate importance to, say, Galileo or Malthus or Rousseau or Mill." One of Wiener's most ardent supporters, Gregory Bateson, called cybernetics "the biggest bite out of the fruit of the Tree of Knowledge that mankind has taken in the past 2,000 years."

Wiener must have found words like that especially gratifying in light of his efforts to position cybernetics as the era's Theory of Everything. Few qualities set Wiener and Shannon apart more than their attitudes toward publicity. "Wiener, in a sense, did a lot to push the idea of cybernetics, which is a somewhat vague idea, and got a lot of worldwide publicity for it," said Stanford's Thomas Kailath. "But that wasn't Shannon's personality at all. Wiener loved the publicity, and Shannon could not have cared less."

The popular success of *Cybernetics* launched a debate over priority within the small clique of mathematicians who wanted to know whether Wiener or Shannon could rightly claim credit for information theory. It also gave rise to a dispute over whether or not Wiener—whose chapter on information as a statistical quantity was admittedly a small section of his book—even knew what information theory meant.

Shannon's 1948 paper, for its part, gave Wiener credit for influencing his view of the statistical nature of communication. But as more attention was paid to the field, Shannon came to realize that he differed from Wiener in some important respects. For one, Shannon insisted that meaning had nothing to do with the transmission of information, a point he believed was critical; Wiener's view of information, on the other hand, included meaning. But perhaps the most significant distinction between their efforts is that the analysis of coding, and its power to protect information transmission from noise, is absent from Wiener's work. Shannon, an engineer by training and inclination, attacked the problem of noise as an engineer might—and his Fundamental Theorem for a Discrete Channel with Noise is the starting point for much of the coding that makes modern information technology possible. This was the key element missing from Wiener's work and the reason, it seems, that Wiener's efforts to claim credit for information theory rankled many of Shannon's followers. As Sergio Verdú, an information theorist of a later era, put it, "in fact, there is no evidence that Wiener ever grasped the notion, at the heart of information theory, of operational meaning lent by a coding theorem."

In the 1950s and 1960s, though, both Shannon and Wiener were more circumspect. Neither took explicit issue with the other's understanding, and though they frequently attended the same conferences and wrote in the pages of the same journals, no significant barbs seem to have been traded between them. But by the 1980s, Shannon concluded that Wiener had not fully comprehended his work. "When I talked to Norbert, like in the 1950s and so on, I never got the feeling that he understood what I was talking about." In another interview, Shannon was even more blunt: "I don't think Wiener had much to do with information theory. He wasn't a big influence on my ideas there, though I once took a course from him." Given Shannon's habitual lack of interest in these sorts of confrontations, statements like those are telling. But for the most part, he left the struggle for credit to others.

By the standards of the great mathematical feuds—Gottfried Leibniz and Isaac Newton battling over custody of calculus, or Henri Poincaré and Bertrand Russell debating the nature of mathematical reasoning—the rivalry between Shannon and Wiener is, sadly, less spectacular than biographers might prefer. But it still stands as an important moment in Shannon's story. Shannon gave the impression of the carefree scholar—someone secure enough in his own intellect and reputation to brush aside the opinion of others. Wiener's opinions and contribution mattered—but not because Shannon worried about who would or wouldn't receive credit. Debates in his field mattered to him less for their opportunities to assert "ownership" of information theory than for their bearing on the substance of information theory itself. Credit, in the end, counted less than accuracy.

20

A Transformative Year

Shannon turned thirty-two in 1948. The conventional wisdom in mathematical circles had long held that thirty is the age by which a young mathematician ought to have accomplished his foremost work; the professional mathematician's fear of aging is not so different from the professional athlete's. "For most people, thirty is simply the dividing line between youth and adulthood," writes John Nash biographer Sylvia Nasar, "but mathematicians consider their calling a young man's game, so thirty signals something far more gloomy." Shannon was two years late by that standard, but he had made it.

Roughly ten years of work had become seventy-seven pages of information theory, and the work had been worthwhile by all accounts. Shannon had won a small measure of fame and had established himself as a first-rate theoretical mind. His work was the springboard for that of others, a sign that he had laid some important foundations. He had made a name for himself within the demanding and insular world of Bell Labs. That year, 1948, would have transformed Shannon on those bases alone. But there was more than mathematics reshaping his life that fall.

Besides jousting with him on matters of the intellect, John Pierce played one other significant role in Claude Shannon's life, in a matter of the heart: he was responsible, indirectly, for introducing Shannon to his future wife, Betty Moore, a young analyst at Bell Labs. Pierce was Moore's immediate supervisor, and it was in the course of dropping by to see Pierce in 1948 that Shannon struck up a conversation with her. Taciturn though he might have been, Shannon had it in him to summon the courage to ask Betty out to dinner. That dinner led to a second, the second to a third, until they were dining together every night.

He charmed her, and they both seemed to share a sense of ironic detachment, a feeling that the world was frequently conspiring to make them chuckle. As their dates grew longer and more frequent, they split time between his West Village apartment and hers on East Eighteenth Street. There, the two shared their mutual loves: mathematics and music. "I played piano and he played clarinet," she recalled, "and we'd come home from work, and we found some books of music that had two parts, and we'd enjoy playing together."

Born on April 14, 1922, Betty was an only child. In the early part of her life, the family lived on Staten Island, but they later moved to Manhattan. Betty Moore's mother and aunt had emigrated to America from Hungary, so the soundtrack of her childhood contained as much Hungarian as accented English. Like many immigrants, the family struggled to find a foothold in their adopted country, and they were struck hard by the Great Depression. Her father faced periods of unemployment, eventually finding a role on the support staff of the *New York Times*. Her mother found steadier work in the fur business, though she was forced to prematurely end her schooling to provide for the family.

Money was always tight. When the Depression hit, they nearly lost their home. A New Deal homeowners' program saved the family from foreclosure, a moment that Betty never forgot. By her daughter's account, "My mother was eternally grateful to FDR and to the New Deal and the protections FDR put in place. They managed to keep the house and survive."

Betty attended Catholic schools, though, she was quick to add, not because of any particular religiosity on her parents' part. Her mother was Catholic, her father Episcopalian, but they chose Catholic schools for Betty because the public school in their neighborhood had unexpectedly closed. Betty proved a gifted student, and by the time she was ready to graduate, several colleges offered her both admission and scholarships.

She had her heart set on Cornell, but the scholarship fell short of the full cost of tuition, and her parents could offer no help. So when a letter arrived offering a full scholarship to the New Jersey College for Women—along with a job offer—Betty wept. Now she'd be able to attend college close to home and even send a bit of money back to her parents. As her daughter, Peggy, recalled, "it changed her life."

Betty Moore studied mathematics at the New Jersey College for Women (now Rutgers University's Douglass College); like many colleges at the time, it was still recovering from the Depression. Enrollments and funding had both been cut, and a feeling of economic uncertainty lingered on campus. Yet by the time Moore was a sophomore, economic troubles had become comparatively trivial. America was at war, and the campus mobilized in support: "students and faculty formed relief organizations, rolled bandages and entered war production industries."

Betty was matter-of-fact, whip-smart, and had a wry sense of humor. She was an avid reader, and those who knew her marked her out as unusually bright. Her choice of major was well suited to the times, and she "fortunately had good grades." As she recalled, "at that point they were very much looking for math majors, particularly women, because the men were all in the service." Bell Labs was among those companies on the search for any and all talented graduates in the field. As she neared graduation day, the Labs gave Betty "the best offer of any of the jobs I had been offered," and she accepted.

She started work in the mathematics department, focusing on microwave research, and then moved to the fast-growing radar group. "Just working there was fascinating," she recalled. And, "considering that the world was in a mess, we were very lucky." She moved back home with her parents and continued to contribute to the household; Betty would support her parents in some way for the rest of their lives.

Of the Claude she knew in the early days, she would say that "he was very quiet and had a wonderful sense of humor." Their courtship began just as Shannon was beginning to achieve a measure of fame for his information theory work, but his rising star seems not to have interfered with their early dating life. That's partly because Shannon was consumed by affection for Betty. It had been seven years since the dissolution of his marriage to Norma, and just as before, the courtship moved efficiently. Betty and Claude met in the fall of 1948 and, by early 1949, Claude had proposed—in his "not very formal" way, as Betty recalled. She accepted, and on March 22, they were married. The wedding was a small affair; as Betty tells it, the only guest "we had from Claude's family was his sister Catherine." The newlyweds soon left the city and moved to Morristown, New Jersey, close to Bell Labs' new facility in Murray Hill.

Nearly all who knew them testified to how good a match Betty was for Claude Shannon—in every sense. It wasn't just the joy he found in her company, though he did. Betty and Claude became professional partners, as well. Albert Einstein famously said of his wife, Mileva Maric, "I need my wife. She solves all the mathematical problems for me." Claude's work was very much his own, but there's no denying Betty's help in bringing it to fruition; she became one of his closest advisers on mathematical matters. She looked up references, took down his thoughts, and, importantly, edited his written work.

Claude's gifts were of the Einsteinian variety: a strong intuitive feel for the dimensions of a problem, with less of a concern for the step-by-step details. As he put it, "I think I'm more visual than symbolic. I try to get a feeling of what's going on. Equations come later." Like Einstein, he needed a sounding board, a role that Betty played perfectly. His colleague David Slepian said, "He didn't know math very deeply. But he could invent whatever he needed." Robert Gallager, another colleague, went a step further: "He had a weird insight. He could see through things. He would say, 'Something like this should be true' . . . and he was usually right. . . . You can't develop an entire field out of whole cloth if you don't have superb intuition."

The trouble with that kind of intuition is that solutions to problems appear before the details and intermediary steps do. Shannon, like many an intuitive mind before him, loathed showing his work. So Betty, who could hold her own mathematically, became his scribe. She was also the first audience for many of his ideas—the most notable exception to the introverted policy of a man who, as she put it, "wouldn't go out of his way to collaborate with other people." Taking dictation, she would also offer her improvements and edits, and add historical references that occurred to her. In later life, when Claude's memory would fail him about this or that reference to a mathematical paper, she would step in and remind him. As Betty put it, "some of his early papers and even later papers are in my handwriting, so called, and not in his, which confused people at first." Confusing, perhaps—but also testament to one of the great mathematical marriages of our time: one that produced path-breaking work and lasted the rest of Claude's life.

21

TMI

Alongside pieces on "Tax Reform," "How to Get a Raise," and "Olin, an Industrial Empire," the December 1953 issue of *Fortune* magazine offered the mass public a first, digestible look at "The Information Theory." Five years after the publication of Shannon's paper in the *Bell System Technical Journal,* it had become the subject of a full-length feature in a magazine whose audience was made up of more than engineers and mathematicians. Francis Bello, *Fortune*'s technology editor and the writer of the piece, was to become one of Shannon's champions in the popular press.

Bello's article opened with a haymaker:

> Great scientific theories, like great symphonies and great novels, are among man's proudest—and rarest—creations. What sets the scientific theory apart from and, in a sense, above the other creations is that it may profoundly and rapidly alter man's view of the world.
>
> In this century man's views, not to say his life, have already been deeply altered by such scientific insights as relatively theory and quantum theory. Within the last five years a new theory has appeared that seems to bear some of the same hallmarks of greatness. The new theory, still almost unknown to the general

> public, goes under either of two names: communication theory or information theory. Whether or not it will ultimately rank with the enduring great is a question now being resolved in a score of major laboratories here and abroad.

Though Shannon praised an early draft of the article, calling it a "bang-up job of scientific reporting," he took characteristic exception to these two opening paragraphs. "Much as I wish it were so, communication theory is not in the same league with relativity and quantum mechanics. The first two paragraphs should be rewritten with a much more modest and realistic view of the importance of the theory." Shannon also urged Bello to acknowledge Norbert Wiener for his contemporary work on cybernetics—and to make sure Bell Labs researchers were given their due.

Bello did give some credit to Wiener and others—but he did nothing to deflate information theory's potential. He continued: "It may be no exaggeration to say that man's progress in peace, and security in war, depend more on fruitful applications of information theory than on physical demonstrations, either in bombs or in power plants, that Einstein's famous equation works."

Comparisons to Einstein were to become a permanent fixture of Shannon's public life. "Shannon is to communications as Einstein is to physics," went a typical line, following Bello's lead. When the town of Gaylord unveiled its Claude Shannon statue, the local paper remembered him as the "Gaylord native son . . . who will be revered forever as the Einstein of the mathematical theory of communication." William Poundstone may have made the most memorable observation: "There were many at Bell Labs and MIT who compared Shannon's insight to Einstein's. Others found that comparison unfair—unfair to Shannon." Despite Shannon's protests, the similarities impressed themselves on his contemporaries: revolutionary theoretical work, a kind of playfulness of spirit, a curious combination of creative skill and the ability to stand apart from the prestige-soaked, ladder-climbing world of elite academia.

But Shannon had to acclimate himself to the praise. In June 1954, shortly after his piece, Bello included Shannon on a list of the twenty most important scientists in America. Beginning with the questions, "What kind of man becomes an outstanding scientist? Is there a widening gulf between him and the rest of society?" Bello interviewed more than 100 scientists and sent questionnaires to dozens more in search of answers.

Along with Shannon, the resulting list included a twenty-six-year-old, boyish-looking molecular biologist working at the Cavendish Laboratory in Cambridge, England. Eight years later, at the age of thirty-four, James Watson won the Nobel Prize, along with Francis Crick and Maurice Wilkins, for discovering the double helix of DNA. Another of Bello's profilees was a thirty-six-year-old wunderkind physicist. Richard Feynman, too, shared a Nobel in 1965 for his work on quantum electrodynamics. In fact, one-quarter of the twenty scientists Bello singled out for recognition would go on to win that honor.

With similar encomia in *Time, Life,* and a host of other major publications, Shannon had reached the heights of scientific celebrity—in a postwar era in which the figure of "the Scientist" had itself reached its apex of cultural prestige.

The press, understandably, was as interested in the curious man behind the new theory of information as it was in the theory itself. Shannon seems to have taken public recognition of his work with a kind of bemused detachment, as in this interview with *Omni*:

OMNI: Did you feel you were destined for fame?

SHANNON: I don't think so. I always thought I was quite sharp scientifically, but scientists by and large don't get the press that politicians or authors or other people do. I thought my paper on switching was quite good, and I got a prize for it, and I thought my information paper was very good, and I got all kinds of acclaim for that—there's a wallful of prizes and stuff in the other room.

OMNI: Do you find fame a burden?

SHANNON: Not too much. I have people like you coming and wasting my afternoons, but that isn't too much of a burden!

By the mid-1950s, Shannon's work had been celebrated in the popular press and applied in a diverse array of fields, sometimes with only the loosest appreciation for what information theory actually meant. For theoretical work as suggestive as information theory—which to a casual reader might appear to offer a rubric for everything from mass media to geology—appropriation and misappropriation were inevitable. For instance: "Birds clearly have the problem of communicating in the presence of noise," ran one contemporary paper. "An examination of birdsong on the basis of information theory might . . . suggest new types of field experiment and analysis." Invoking "information theory," like any fashionable term, was often a shortcut to research funding. At the same time, the elegance and simplicity of Shannon's theory made it an attractive tool across disciplines. Even if the potential for overuse had troubled him, the normally conflict-averse Shannon might have been expected to simply laugh, shrug his shoulders, and move on to other problems. In the main, this is mostly what he did—with one important exception.

In 1955, Louis A. de Rosa, the chairman of the Institute of Radio Engineers' Professional Group on Information Theory, published an editorial in the group's newsletter. De Rosa's "In Which Fields Do We Graze?" was a genuine query of his colleagues working in information theory:

> The expansion of the applications of Information Theory to fields other than radio and wired communications has been so rapid that oftentimes the bounds within which the Professional Group interests lie are questioned. . . . Should an attempt be made to extend our interests to such fields as management, biology, psychology, and linguistic theory, or should the concentration be strictly in the direction of communication by radio or wire?

Shannon himself took to the pages of IRE's journal to address the matter in a brief pronouncement titled "The Bandwagon." The

573-word response begins: "Information theory has, in the last few years, become something of a scientific bandwagon. Starting as a technical tool for the communication engineer, it has received an extraordinary amount of publicity in the popular as well as the scientific press." Shannon allowed that the popularity was, at least in part, due to information theory's hovering place on the edges of so many of the era's hottest fields—"computing machines, cybernetics, and automation"—as well as to its sheer novelty.

And yet, he continued, "it has perhaps been ballooned to an importance beyond its actual accomplishments. Our fellow scientists in many different fields, attracted by the fanfare and by the new avenues opened to scientific analysis, are using these ideas in their own problems. . . . In short, information theory is currently partaking of a somewhat heady draught of general popularity." Shannon was willing to concede that all of the momentary attention was "pleasant and exciting." Still,

> it carries at the same time an element of danger. While we feel that information theory is indeed a valuable tool in providing fundamental insights into the nature of communication problems and will continue to grow in importance, it is certainly no panacea for the communication engineer or, *a fortiori,* for anyone else. Seldom do more than a few of nature's secrets give way at one time.

Seldom do more than a few of nature's secrets give way at one time. It's a remarkable statement from someone who still had a full career ahead of him, someone who, in a practical sense, had every incentive to encourage information theory's inflation. Yet here was Shannon pulling on the reins. He continued: "It will be all too easy for our somewhat artificial prosperity to collapse overnight when it is realized that the use of a few exciting words like information, entropy, redundancy, do not solve all our problems."

In place of all this feverish excitement, Shannon counseled moderation:

> Workers in other fields should realize that the basic results of the subject are aimed in a very specific direction, a direction that is not necessarily relevant to such fields as psychology, economics, and other social sciences. Indeed, the hard core of information theory is, essentially, a branch of mathematics, a strictly deductive system. . . . I personally believe that many of the concepts of information theory will prove useful in these other fields—and, indeed, some results are already quite promising—but the establishing of such applications is not a trivial matter of translating words to a new domain, but rather the slow tedious process of hypothesis and experimental verification.

Above all, he advised his colleagues that

> we must keep our own house in first class order. The subject of information theory has certainly been sold, if not oversold. We should now turn our attention to the business of research and development at the highest scientific plane we can maintain. Research rather than exposition is the keynote, and our critical thresholds should be raised. Authors should submit only their best efforts, and these only after careful criticism by themselves and their colleagues. A few first rate research papers are preferable to a large number that are poorly conceived or half-finished. The latter are no credit to their writers and a waste of time to their reader.

The editorial, and others that agreed with Shannon's position, had the intended effect. Robert Gallager offered this observation about Shannon's approach to conflict: "Claude Shannon was a very gentle person who believed in each person's right to follow his or her own path. If someone said something particularly foolish in a conversation, Shannon had a talent for making a reasonable reply without making the person appear foolish." Given that habitual restraint, the "Bandwagon" editorial was a telling statement. That he was moved to write such a piece showed his true depth of concern over the use and abuse of information theory—and his worries that, instead of birthing a new field of science, he had only inflated a speculative bubble.

Betty Shannon confessed that Shannon was perhaps more frustrated than even the editorial let on: "He got a little irritated with the way people were pulling it around. People didn't understand what he was trying to do." Robert Fano would go further, citing his own frustration as well as Shannon's: "I didn't like the term Information Theory. Claude didn't like it either. You see, the term 'information theory' suggests that it is a theory about information—but it's not. It's the transmission of information, not information. Lots of people just didn't understand this."

For Shannon, useful, informed applications of information theory were always welcome. But claims for its hyperimportance—attempts to position it as the century's Key to All Mythologies—invariably rested on the kinds of airy generalities and lazy philosophizing that he scorned. Here was the real danger: that the ideas he had set in motion might become so diffuse so as to lose all meaning. That danger is, perhaps, a risk inherent to any revolution in scientific thought. But Shannon felt compelled to do his part to ward it off. His work had opened a theoretical and metaphorical Pandora's box. "The Bandwagon" was his attempt to shut the lid, discipline the discipline, and remind at least the engineering world that the theory he had pioneered—and the work that had made him famous—could only remain meaningful within its own proper bounds.

22

"We Urgently Need the Assistance of Dr. Claude E. Shannon"

Dear Dr. Kelly," the letter began, "although I am well aware of the patriotic contribution which you and your Company are already making in solving the many problems presented to you by the United States Government, I must make a personal request of you in a matter of the most urgent concern and importance to the security of the United States." Typed on official Central Intelligence Agency letterhead and delivered to the head of Bell Labs, the message was deliberately vague:

> In attempting to find a solution to an especially vital problem confronting us at this time, we urgently need the assistance of Dr. Claude E. Shannon of your Company who, we are informed on the best authority, is the most eminently qualified scientist in the particular field concerned. . . . If his services could be made available for this purpose on a basis satisfactory to both you and Dr. Shannon, I will be deeply grateful. I fully realize that even his temporary absence will be a great inconvenience to your organization and you may be sure that only the most compelling reason would cause me to make this request.

The writer of the letter was one of the most distinguished military men of his era: Walter Bedell Smith, the CIA director, former Army chief of staff to Dwight Eisenhower, and former ambassador to the Soviet Union. He was also the fourth person to lead the CIA, a job that, at that time, held little of the public profile it does now. Three days later, a copy of the same letter was sent from Kingman Douglass to Captain Joseph Wenger of the U.S. Navy, with a small attachment. "I hope very much this letter will succeed in its purpose." Shannon's past work offers some indication of what the CIA was after, but the fact that Douglass and Wenger were involved makes it even clearer.

Kingman Douglass was one of the sons of the upper crust whose life was a mix of prestigious private schools, paneled boardrooms, and pressurized war rooms. A graduate of the Hill School and Yale University, he flew planes in World War I and ran intelligence operations in World War II. He also served on two separate occasions with the CIA, including as assistant director for current intelligence.

Joseph Wenger also spent his career in the highest echelons of the intelligence world. "One of the first naval officers to realize the role of communications intelligence," he was a U.S. Naval Academy graduate who would rise to become a rear admiral—and along the way would transform the way the Navy thought about and implemented cryptologic operations, becoming "one of the architects of centralized cryptology." In the Pacific Theater of World War II, he found that the close study of Japan's "message externals," or seemingly trivial details ranging from call signs to communication habits, could be as cryptographically fruitful as the analysis of the messages themselves. By 1949, with two wars' worth of experience and understanding, he was a leader in the Armed Forces Security Agency (AFSA), the forerunner to the modern NSA.

In a call with Shannon, Wenger made the case that the intelligence community needed his help. Wenger relayed the results to Douglass in a barely legible note: "I spoke to Shannon today on the phone and he appeared open to persuasion. He said he would reserve judgment until he could learn more of the problem and determine whether or not it is something he felt he could contribute. I offered to send an emissary to explain better to him as soon as his clearance is O.K." John von Neumann also contacted Shannon that week, impressing upon him the significance of the request. It was very much in keeping with Shannon's sensibility that he was neither overawed by being sought out for such a consultancy nor too quick to jump at the problem before knowing its full scope.

A week after the letter from CIA director Smith, Wenger and Douglas received a response from Bell Labs' Mervin Kelly:

> While there have been several other approaches to enlist Dr. Shannon's services in connection with military activities and it has been our judgment that, in general, he could best contribute in his particular field by carrying on his researches independently, the matter with which your letter deals is of a more compelling nature and we shall, therefore, be glad to encourage and assist Dr. Shannon in participating to the extent of the preliminary examination you suggest.

This note sums up Shannon's life in the early 1950s. Applications of information theory had mushroomed. The demands on his time had multiplied, and Shannon was doing his level best to keep the hordes at bay. When his resistance failed, it was almost always because of forces outside his control. Shannon made a principle of indifference; it was central to a career in which he chased his instincts, often at the expense of more prestigious or remunerative options. But his work on information theory had brought him national renown. And now, the federal government was asking for him by name.

The war's end had brought the military a thorny problem: the exit from public service of many of the nation's top scientists, mathematicians, and engineers. Beginning in wartime, as Sylvia Nasar wrote, "to be plucked from academe and initiated into the secret world of the military had become something of a rite of passage for the mathematical elite." Now, though, "how to keep the best and brightest thinking about military problems was far from obvious. Men of the caliber of John von Neumann would hardly sign up for the civil service." One solution, familiar to the men who occupied the upper rungs of the mathematical world, was the establishment of technical committees in close contact with various branches of the defense establishment. The committee that would become most familiar to Shannon—and the reason for the urgent messages from Wenger and von Neumann—was known as the Special Cryptologic Advisory Group, or SCAG.

In the NSA's words, "the fundamental purpose in establishing SCAG was to assemble a specific group of outstanding technical consultants in the scientific fields of interest to the Agency, and thus provide a valuable source of advice and assistance in solving special problems in the cryptologic field." Like most groups of this kind, SCAG was a means to a host of ends. There were knotty technical problems on which real, practical advice was sought. Committee members served as de facto headhunters, finding and sourcing talent at the request of senior government officials. There were frank exchanges about the nation's readiness on a number of fronts. The first meeting of SCAG included sessions on the value and importance of communications intelligence, on a case study of a complex intelligence problem from World War II, on the state and purpose of the intelligence bureaucracy itself, and on an AFSA project code-named SWEATER. The committee's concern ran the gamut from the technical to the philosophical.

From the time he was asked to serve in 1951 until the mid-1950s, Shannon made regular trips to Washington, D.C., for these meetings, serving on SCAG and its successor committee, the National Security Scientific Advisory Board. These meetings were multi-day affairs, and each day featured dawn-to-dusk sessions with the nation's top brass discussing their most urgent intelligence dilemmas. "Because a considerable portion of each agenda had to be devoted to briefings by NSA officials before the Board could get to the consideration of Agency problems, it was of the utmost importance that the agenda contain the problems thought to be most pressing." There was another practical reason that only the most pressing problems were brought to SCAG: these were men whose schedules were notoriously hard to align. In fact, a substantial measure of the record available to us about SCAG and other such committees concerns the challenge of herding about a dozen men of the highest scientific accomplishment into one room at the same time.

Boards of this kind were, by design, hamstrung. In the NSA

historian's own words, "Lacking accessible, secure areas, some of the advisors were handicapped by their inability to hold and consult cryptologic documents between meetings, and thus to live more or less with a problem. They could not benefit from the intuitive concepts that come with prolonged, even if intermittent, attention." But the boards at least served the purpose of keeping the NSA leadership broadly connected with the scientific world.

The leadership with which Shannon was interacting had come of age in the midst of two massive intelligence failures. The horror of Pearl Harbor was seared in their memories. More recently, the invasion of South Korea by North Korea had again blindsided American policy makers, and by 1950, the country was again on a war footing. Which is all to say that Shannon was speaking with and working for men who had seen armed combat and were sending a new generation of Americans into another bloody conflict. The stakes were real; the intelligence requirements were manifold. Mathematical thinkers of Shannon's and Von Neumann's caliber were a necessary external measure of the defense establishment's technological and scientific soundness.

It's only from recently declassified documents that we get even this vague sense of Shannon's work for the government during this time, and still many of the salient details remain classified. Shannon himself was cagey about what he did. Decades later, in an interview with Robert Price, he ducked the questions:

> **PRICE:** And you were on the board of the NSA, weren't you for a while?
>
> **SHANNON:** I don't think I was on the board. I might have been a member. I don't think I was that . . . elevated a position.
>
> **PRICE:** Well, you had dealings with the National Security Agency at some time I've been told.
>
> **SHANNON:** Yes, that's a better way to put it. . . . I got involved in cryptography at a later period. I was a consultant. I probably should . . . I don't know . . .

PRICE: You're talking about NSA now, probably the Advisory Board?

SHANNON: Well, I was invited . . . I think, I don't know that I have any . . . even though this was a long time ago, I'd better not talk about. . . .

To some extent, this is classic Shannon: immune to self-puffery, unwilling to dive into topics that held limited interest. But Shannon tended to parry questions of this sort with a wry mix of sarcasm and humor. That he was nervous and halting in this exchange says a great deal about the secrecy surrounding the work he did.

Shannon had reason to be guarded: he had exposure to some of the nation's most closely held secrets and systems and came into contact with the founding fathers and texts of the national security state. He understood both the gravity of the work and the need to keep the privileged information secret. This was no idle matter. One of Shannon's fellow NSA scientific advisers, John von Neumann, was watched round-the-clock by uniformed military personnel when he was on his deathbed at the Army's Walter Reed Hospital. Impressive though von Neumann's mind may have been, it wasn't immune from infiltration—or so the government feared. And what better time to infiltrate it—and grab the precious state secrets it held—than when it was in a medically induced haze?

23

The Man-Machines

Could a machine think?—Could it be in pain?—Well, is the human body to be called such a machine? It surely comes as close as possible to being such a machine. But a machine surely cannot think!—Is that an empirical statement? No. We only say of a human being and what is like one that it thinks. We also say it of dolls and no doubt of spirits too.

—Ludwig Wittgenstein

I'm a machine and you're a machine, and we both think, don't we?

—Claude Shannon

If Shannon had peculiar work habits before the publication of his information theory, his growing reputation granted him the license to indulge those peculiarities without reservation. After 1948, the Bell Labs bureaucracy could not touch him—which was precisely as Shannon preferred it. Henry Pollak, director of Bell Labs' Mathematics Division, spoke for a generation of Bell leaders when he declared that Shannon "had earned the right to be non-productive." Shannon arrived at the Murray Hill office late, if at all, and often spent the day absorbed in games of chess and hex in the common areas. When not besting his colleagues in board games, he could be found piloting a unicycle through Bell Labs' narrow passageways, occasionally while juggling; sometimes he would pogo-stick his way around the Bell Labs campus, much to the consternation, we imagine, of the people who signed his paychecks.

Colleagues may have bristled at all this, but Shannon was, by this point, a legend masquerading as an ordinary employee. He had engineered as close to an emeritus role for himself as he could under the contractual obligations of full-time employment. That meant being able to work with his door closed, practically a sin in Bell Labs circles. It also meant pursuing personal projects to whatever conclusion suited him. One receipt from this time records a series of hardware store purchases that he billed to the Labs, presumably part of a machine Shannon was building, the result of which could hardly have mattered to the practical work of the phone company.

But none of this was cause for alarm within Bell Labs. There was no doubting the quality of Shannon's mind, and thus no one thought to ask rigorous questions about how he was keeping it occupied. After all, the "founder of information theory" had, essentially, dropped the theory into everyone's laps after finishing it in private. Who was to question what *else* he might be up to behind closed doors?

One curious side effect of all this freedom: Shannon became, during this period, an inconsistent correspondent, even as the volume of his correspondence began to grow with his reputation. Letters went unanswered for long stretches, so many and for so long that Shannon collected them all in a folder labeled "Letters I've procrastinated in answering for too long." In the words of Jon Gertner, "it seemed lost on Shannon that the scientist who had declared that any message could be sent through any noisy channel with almost perfect fidelity was now himself a proven exception. Transmissions could reach Claude Shannon. But then they would fail to go any further."

These weren't all anonymous fans and unknown admirers, either. Shannon received correspondence from eminent scientists, high-level government officials—and even L. Ron Hubbard.

It's only with the benefit of several decades that we can judge this to be one of the stranger correspondences in Shannon's life. It's worth being emphatic on one point: yes, the founder of dianetics and the

Church of Scientology sought Shannon out. No, Shannon was not a Scientologist himself. Hubbard, it appears, was more interested in Shannon than Shannon was in Hubbard. Shannon, for his part, wrote a letter to Warren McCulloch, a leading cybernetics researcher at MIT, asking if McCulloch would meet with his "friend" Hubbard.

Hubbard, it seems, was better known to Shannon as an author of space operas than as a budding religious crackpot. "If you read Science Fiction as avidly as I do you'll recognize him as one of the best writers in that field," Shannon writes. "Hubbard is also an expert hypnotist and has been doing some very interesting work lately in using a modified hypnotic technique for therapeutic purposes. . . . I am sure you'll find Ron a very interesting person, with a career about as varied as your own, whether or not his treatment contains anything of value."

Hubbard would later write to Shannon thanking him for help on his research and promising a copy of *Dianetics* when it was released. No further correspondence between the founder of information theory and the pope of Scientology has been recorded. Yet, as William Poundstone notes, "to this day Hubbard's Scientology faith cites Shannon and information theoretic jargon in its literature and web sites."

The correspondence with Hubbard was positively staid in comparison with many of the other letters that piled up on Shannon's desk. Alongside the usual traffic of scientific colleagues reaching out for a paper or a book review, there was also a steady stream of cranks who sought out Shannon's approval for their private researches, or whose paranoia about the phone company led them to contact one of its prominent faces. One handwritten letter began, "Dear Dr. Shannon, I am enclosing a 'Theory of Space.' I have sent it to several other eminent scientists but thus far have not received a reply. . . . " A letter from the self-identified "IDEA MAN" requested Shannon's assistance "to complete and verify a 15 year search to locate and exactness OF life, mind, and energy."

Another was more menacing:

> Dear Sir, Your mechanical robot Bel, the idol in the Bible, is a mechanical monstrosity. Your robot is breaking five amendments

> of the Constitution (1, 3, 4, 5, + 13th). God admits that I am laughing at you. You are making a traitor out of the President of the U.S. And the F.B.I. By letting your robot deceive you. I have threatened to sue the NY Telephone Co. Of NY City, and I will, if you don't wake up.

Shannon was good-natured about it all—able to use his charm to defuse tough inquiries, or more often, to ignore them without a second thought. Unlike many scientists who parlayed successful research careers into lives as public intellectuals, he did not seem to consider using his growing standing within the world of science as an opportunity to expand his network outside of it. Nor did he take it as a responsibility to opine on policy or act as a public educator. If anything, he closed himself off further, ignoring letters, colleagues, and projects, and spending his time and attention absorbed by the puzzles that interested him most. Shannon had earned this right—information theory was painstaking work—and he found himself drawn now to new problems and fresh horizons, including some that seemed, to colleagues, borderline ridiculous for someone of Shannon's stature.

"I think the history of science has shown that valuable consequences often proliferate from simple curiosity," Shannon once remarked. Curiosity in extremis runs the risk of becoming dilettantism, a tendency to sample everything and finish nothing. But Shannon's curiosity was different. His kind meant asking a question and then constructing—usually, with his hands—a plausible answer. Could a robotic mouse navigate its way through a maze? Build one and find out. Could a machine turn itself off? Make one that is trained to commit technological hara-kiri. What other people called hobbies, he thought of as experiments: exercises in the practice of simplification, models that filed a problem down to its barest interesting form. He was so convinced of a machine-enabled future, and so eager to explore its boundaries, that he was willing to tolerate a degree of ridicule to bring it to pass. He was preoccupied, as he wrote to a correspondent, "with the possible capabilities and applications of large scale electronic computers." Considered in the light of that future, our present, his machines weren't hobbies—they were proofs.

One starting point for all of this tinkering with mechanics was benign: a wife's Christmas gift. "I went out and got him the biggest Erector set you could buy in America. It was fifty bucks and everyone thought I was insane!" Betty Shannon later told an interviewer. Claude Shannon added: "Giving it to a grown man! But the fact of the matter is that it was extremely useful and I used it to try out different things." And like a child with a new present, Shannon became obsessed: the basement became a mess of loose erector parts, and he stayed up late into the night, building away.

The first idea was a dry run: a mechanical turtle that stalked the Shannon house, bumping into the walls and turning around only to bump into a different wall. But the hapless turtle anticipated the next invention, the one that would, unexpectedly, attract national attention: Theseus, the maze-solving mouse. As one report had it, the idea for a mechanical mouse that could navigate a maze grew out of Shannon's attempt to escape the famous garden maze at London's Hampton Court Palace. It took him twenty minutes; he figured it could be done faster. Later, the most famous photo of Claude Shannon would depict him with a finished Theseus and the maze, Shannon's hand setting the mouse down within the walls. It was named, somewhat optimistically, for the Greek hero who slew the Minotaur and escaped the fearsome Labyrinth; but for now it was a three-inch piece of wood with copper whiskers and three wheels.

It was Shannon's research on switching, and his work for the telephone company, that inspired the guts of the contraption. Seventy-five electromechanical relays, the sort used as switches in the phone system to connect one call to another, toggled like railroad tracks shifting trains to allow the mouse to navigate. Betty completed the wiring for the earliest prototype. "We did all this at home at night after work," she said.

Theseus was propelled by a pair of magnets, one embedded in its hollow core, and one moving freely beneath the maze. The mouse would begin its course, bump into a wall, sense that it had hit an obstacle with its "whiskers," activate the right relay to attempt a new path, and then repeat the process until it hit its goal, a metallic piece of cheese. The relays stored the directions of the right path in "memory": once the mouse had successfully navigated the maze by trial and error, it could find the cheese a second time with ease. Appearances to the contrary, Theseus the mouse was mainly the passive part of the endeavor: the underlying maze itself held the information and propelled Theseus with its magnet. Technically, as Shannon would point out, the mouse wasn't solving the maze; the maze was solving the mouse. Yet, one way or another, the system was able to learn.

When it arrived at Murray Hill, Theseus became a minor Bell Labs celebrity. The mouse earned Shannon and the Labs a patent. The Labs also commissioned a short film starring Shannon and Theseus. The seven-minute short was produced with the general public in mind. Shannon, nattily dressed in a dark suit with a light red tie, explains the maze-solving mouse and its mechanisms in the deliberate, step-by-step manner of a college professor. "Hello," he begins, "I'm Claude Shannon, a mathematician here at the Bell Laboratories." He dives in, explaining both what the viewer is watching—the mouse making its way through a maze—and what underlies the system. When it comes to extending the analogy of the mouse and maze further than what's in front of him, Shannon only hints at it, only gestures in the direction of what Theseus means for the possibilities of a robot brain:

> Of course, solving a problem and remembering the solution involves a certain level of mental activity, something akin, perhaps, to a brain. A small computing machine serves Theseus for a brain. . . . We have placed the brain cells of Theseus, if you like, behind a small mirror here.

Shannon explains that Theseus's brain is something more basic and familiar, something akin to the system that powers the telephone's elaborate network of switches and wires. "Here at the Bell Telephone Laboratories, we're concerned with improving your telephone system," Shannon says, coming the closest he ever will come to shilling for his employer. That moment, along with the images of telephones dialing and switches activating, and the cheery music in the background, was a necessary piece of PR: concerned as they were about regulatory interest in their work, the higher-ups at Bell Labs and AT&T couldn't allow Claude Shannon to go into theaters, schools, or universities with a robotic mouse and risk giving the appearance that the enormous leeway and profits they had been granted by the U.S. government was being devoted to frivolities.

Shannon closes the video by changing the contours of the maze and putting Theseus inside a square with no exit. "Like the rest of us, he occasionally finds himself in a situation like this," Shannon says, as the mouse moves, hits a wall, moves, hits a wall, and ends up trapped. The camera cuts to Shannon, who smirks, and the music cues the end of the demonstration.

The outside world took an unexpected interest in Theseus, and the celebrity it earned Shannon and Bell Labs impressed Shannon's bosses. One story lodged itself in Bell Labs lore. Henry Pollak recounted what happened when Shannon gave a demonstration of Theseus for the AT&T board of directors:

> I was told that one of the board members at the end of the presentation said, "Now *that's* the kind of original thinking we need at AT&T! I propose that Claude Shannon be made a member of the board!" And they had a hell of a time dissuading this guy from the idea of making Claude Shannon a board member, and they finally got around it by the fact that Claude Shannon did not own enough shares of stock to be made a board member.

Time magazine featured Theseus in a short article: "Mouse with a Memory." *Life* published a photo of Theseus finding the cheese. *Popular Science* ran a three-page spread under the headline "This Mouse Is Smarter Than You Are." Theseus found its way into more serious quarters, as well. The mechanical mouse was a featured subject of discussion at the famed 1951 Macy Conference, an interdisciplinary meeting of scientists and scholars in New York. Shannon was in attendance along with many of the leading authorities on artificial intelligence and computing, as well as the anthropologist Margaret Mead. The incongruity of such leading minds discussing a mechanical mouse was mitigated by the fact that Theseus (or, to be exact, the mouse-maze system as a whole) was a working example of the "artificial intelligence" that many of the esteemed attendees had spent their careers pondering only in theory. Theseus *was* artificially intelligent. When an attendee pointed out the obvious—that if the metallic cheese were removed, the mouse would simply sputter along, searching in vain for a piece of cheese that was no longer there—conference attendee and social scientist Larry Frank responded, "It is all too human."

In the end, the editors of the conference proceedings offered a skeptical assessment of Theseus (in the process demoting him, perhaps unconsciously, from "mouse" to "rat"):

> The fascination of watching Shannon's innocent rat negotiate its maze does not derive from any obvious similarity between the machine and a real rat; they are, in fact, rather dissimilar. The mechanics, however, is strikingly similar to the *notions* held by certain learning theorists about rats and about organisms in general.

In other words, Theseus was not a real intelligence, but he did model one aspect of how a rat or any other another creature might learn. If Shannon indulged in a good-natured eye roll, it is not recorded.

Shannon would later tell a former teacher of his that Theseus had been "a demonstration device to make vivid the ability of a machine to solve, by trial and error, a problem, and remember the solution." To the question of whether a certain rough kind of intelligence could be "created," Shannon had offered an answer: yes, it could. Machines could learn. They could, in the circumscribed way Shannon had demonstrated, make mistakes, discover alternatives, and avoid the same missteps again. Learning and memory could be programmed and plotted, the script written into a device that looked, from a certain perspective, like an extremely simple precursor of a brain. The idea that machines could imitate humans was nothing new. But Theseus had made that idea—and the promise that a machine could memorize and deduce—seem vividly real.

Over the years, Shannon's thinking and nonthinking machines took on a range of shapes and styles. Some served as an oblique social commentary: the "Ultimate Machine," when its single switch was flipped, would reach out a mechanical hand and turn itself off. THROBAC ("Thrifty Roman-Numeral Backward-Looking Computer") was a calculator whose keys, processing, and output all worked in Roman numerals, useless except to those who could decipher the difference between, say, CLXII and CXLII. These gadgets had the character of sly, private jokes. But Shannon also placed a high value on his tinkering. "The design of game playing machines may seem at first an entertaining pastime rather than a serious scientific study," he allowed, but there was "a serious side and significant purpose to such work, and at least four or five universities and research laboratories have instituted projects along this line."

His goals were as grand as the means, at least at the time, were simple. "My fondest dream is to someday build a machine that really thinks, learns, communicates with humans and manipulates its environment in a fairly sophisticated way," Shannon admitted. But he was not bothered by the usual fears of a world run by machines or a human race taking a backseat to robots. If anything, Shannon believed the opposite: "In the long run [the machines] will be a boon to humanity, and the point is to make them so as rapidly as possible. . . . There is much greater empathy between man and machines [today] . . . we'd like to close it up so that we are actually talking back and forth."

That quote, and several of the anecdotes that followed Shannon until the end of his life, originated in a now largely forgotten profile of Shannon in *Vogue* magazine, titled "The Man-Machines May Talk First to Dr. Shannon." As a part of the profile, Shannon spoke at length with writer Brock Brewer about the connection between automata and their creators. (And as it was a feature for *Vogue,* and not, say, *Scientific American,* Shannon was expected to endure a photo shoot, with the renowned Henri Cartier-Bresson behind the lens. It put Shannon in illustrious company: Cartier-Bresson's other shoots included Mahatma Gandhi's funeral, Queen Elizabeth's coronation, and the first several months of Mao Zedong's ascendance.)

The piece opened with what, at the time, must have seemed the musings of a madman: "Dr. Claude E. Shannon . . . who creates, plays with, stays a think ahead of thinking machines, looks forward to man and machine talking back and forth. For him, why not?" For Shannon, the prospect of artificial intelligence was a tangible reality, not a futuristic fantasy. Imagining how "computer-controlled exploratory robots" would handle accidentally falling into a hole on the moon (and anticipating the Roomba in the process), he said,

> you have to think of problems like this when machines are running around loose in the real world. A machine on the moon must protect itself—not fall down a hole, without your having to tell it not to. It's the same problem we're going to have some day with furniture when there are robot housekeepers running around the house, picking up things.

Shannon was happily oblivious to fears of exponentially expanding artificial intelligence, of robots begetting ever-more-advanced robots and putting the human race at risk. In fact, his was a thoroughly optimistic vision of technological progress—one in which machines ought to be given increasing abilities, responsibilities, and information. In response to the question of what the point of all his robot work might be, Shannon remarked that his goals were threefold: "First, how can we give computers a better sensory knowledge of the real world? Second, how can they better tell us what they know, besides printing out the information? And third, how can we get them to react upon the real world?"

Or, as he told a later interviewer, in an even more optimistic mood:

> I believe that today, that we are going to invent something, it's not going to be the biological process of evolution anymore, it's going to be the inventive process whereby we invent machines which are smarter than we are and so we're no longer useful, not

> only smarter but they last longer and have replaceable parts and they're so much better. There are so many of these things about the human system, it's just terrible. The only thing surgeons can do to help you basically is to cut something out of you. They don't cut it out and put something better in, or a new part in.

In fact, when it came to human superiority over machines, "thinking is sort of the last thing to be putting up a fight." While Shannon did not expect a computer to pass the famous, and famously open-ended, Turing Test—a machine indistinguishably mimicking a human—within his lifetime, in 1984 he did propose a set of more discrete goals for artificial intelligence. Computer scientists might, by 2001, hope to have created a chess-playing program that was crowned world champion, a poetry program that had a piece accepted by the *New Yorker,* a mathematical program that proved the elusive Riemann hypothesis, and, "most important," a stock-picking program that outperformed the prime rate by 50 percent. He was about half right: a computer did defeat the world chess champion in 1997, four years before Shannon's deadline, and computers do conduct the bulk of the world's stock trading.

Yet there were moods in which Shannon's cheeriness over the future of machines curdled into misanthropy. "We artificial intelligence people are insatiable," he once wrote. Once machines were beating our grandmasters, writing our poetry, completing our mathematical proofs, and managing our money, we would, Shannon observed only half-jokingly, be primed for extinction. "These goals could mark the beginning of a phase-out of the stupid, entropy-increasing, and militant human race in favor of a more logical, energy conserving, and friendly species—the computer."

24

The Game of Kings

The crowd at Philadelphia's Masonic Hall may have heard rumors of the mysterious machine that had the ability to play chess, but the day after Christmas of 1826 was, for most, the first time they would see the "Chess Playing Automaton" in the flesh. Johann Maelzel, a showman par excellence, took the stage and directed the attention of the audience to the machine by his side: a box about the size of an executive desk, with a mannequin emerging from the top, dressed in robes and a turban in the style of "an oriental sorcerer."

With theatrical flourish, Maelzel opened the side door to "The Turk," revealing gears and gadgetry. By the time the machine had checkmated its first opponent, the crowd was astounded. Silas Weir Mitchell, the eminent physician and writer, was moved to observer that "the Turk—even he, with his oriental silence and rolling eyes, would haunt your nightly visions for many an evening after. Since then we have known him better, and we confess to this day a certain mysterious awe of his eternal cross-leggedness, his turbaned front, and left-handed activity." In that face of that, sorcery seemed a plausible explanation.

The only sorcery, though, was how Maelzel managed to get away with it. The Turk was a hoax: inside the elaborately designed device, behind the gears and pulleys, was a human chess player, navigating each game like a puppeteer. Some of the era's greatest chess players would power the Turk, and yet its secret remained hidden from the public for several decades. Edgar Allan Poe, among others, was prescient enough to investigate the phenomenon, fixing his suspicion on one of the Turk's handlers "who is never to be seen during the exhibition of the Chess Player, although frequently visible just before and just after."

But Poe's skepticism was a minority opinion, and for much of the century it was believed that the machine was just as good—and terrifying—as advertised. The Turk tapped into an anxiety as persistent as it was powerful. Before the legend of John Henry and the fear of a machine to surpass man's might, before science fiction imagined artificial intelligence or the singularity, there was the Turk, a machine that claimed to surpass its creators. Of course, the Turk was a hoax; but the hoax was only a reprieve.

If Shannon was cheerfully optimistic about the possibilities of thinking machines, it wasn't only because he had built a mechanized mouse that could find its way through a maze to a piece of steel cheese—and remember its path. It was also because, in the late 1940s and early 1950s, he turned his curiosity to the question of how and whether a computer might be programmed to compete against a human being in a chess match. It didn't matter that the recent history of such machines was a story of hucksters; Shannon believed it was possible for a computer to play honestly, and to play better than a human. What this research gave Shannon was yet more assurance that a properly programmed machine could do more than mimic a human brain—it could best it.

In a life of pursuits adopted and discarded with the ebb and flow of Shannon's promiscuous curiosity, chess remained one of his few lifelong pastimes. One story has it that Shannon played so much chess at Bell Labs that "at least one supervisor became somewhat worried." He had a gift for the game, and as word of his talent spread throughout the Labs, many would try their hand at beating him. "Most of us didn't play more than once against him," recalled Brockway McMillan.

On a trip to Russia in 1965, Shannon offered a friendly game to

Soviet international grandmaster and three-time world champion Mikhail Botvinnik. Botvinnik, having presumably endured countless games of show for various dignitaries, agreed to the match but played without paying much attention and nursed a cigarette throughout, his uninterest apparent to all in the room. Then, suddenly, Shannon managed to win the favorable exchange of his knight and a pawn for Botvinnik's rook early in the contest. Botvinnik's attention was instantly yanked back to the board, and the atmosphere of the room shifted as the Russian champion realized that his challenger was more than just another hapless dignitary. "Botvinnik was worried," Betty would remember years later.

The game went on far longer than anyone, including the surprised champion, could have predicted. But there was still no real doubt about the outcome. After forty-two moves, Shannon tipped his king over, conceding the match. Still, lasting dozens of moves against Botvinnik, considered among the most gifted chess players of all time, earned Shannon lifelong bragging rights.

(Another incident from the same trip to Russia spoke to his and Betty's sense of humor. When Shannon complained aloud that the lock on their hotel room's door was broken, a locksmith instantly appeared—leading them to suspect that their room had been bugged by the Soviet authorities. Their next move was to complain aloud that they had never received the royalties for the Russian edition of his book—and a check materialized the next day.)

His work on computerized chess would, in time, be recognized as another instance of Shannon dropping into a field and, in one stroke, defining its limits and unearthing many of its central possibilities. Decades after the publication of his paper "Programming a Computer for Playing Chess," *Byte* magazine would put it succinctly: "There have been few new ideas in computer chess since Claude Shannon." The paper that would bring the world a significant step closer to an actual, working Turk attracted none of the hoax's audience or attention. Shannon introduced his idea for a chess-playing computer with characteristic modesty: "Although perhaps of no practical importance, the question is of theoretical interest, and it is hoped that a satisfactory solution of this problem will act as a wedge in attacking other problems of a similar nature and of greater significance."

Shannon imagined some future applications of a chess-playing artificial intelligence: routing phone calls, translating text, composing melodies. As he reminded his readers, these machines were right around the technological corner, and no one doubted their economic utility. As diverse as these applications were, they had an important quality in common: they didn't operate according to a "strict, unalterable computing process." Rather, "solutions of these problems are not merely right or wrong but have a continuous range of 'quality.'" In this way, chess was a valuable test case for the emerging generation of artificial intelligence.

Nearly a half century before Deep Blue defeated the world's human champion, Shannon anticipated the value of chess as a sort of training ground for intelligent machines and their makers:

> The chess machine is an ideal one to start with, since: (1) the problem is sharply defined both in allowed operations (the moves) and in the ultimate goal (checkmate); (2) it is neither so simple as to be trivial nor too difficult for satisfactory solution; (3) chess is generally considered to require "thinking" for skillful play; a solution of this problem will force us either to admit the possibility of a mechanized thinking or to further restrict our concept of "thinking"; (4) the discrete structure of chess fits well into the digital nature of modern computers.

Shannon believed that, at least within the realm of chess, the inanimate had certain intrinsic advantages. The obvious ones were processing speeds well beyond the human brain and an endless capacity for computation. Further, an artificial intelligence wouldn't be susceptible to boredom or exhaustion; it could continue to drill into a chess position well after its human counterpart had lost concentration. Computers were, in Shannon's view, blessed with "freedom from errors," their only mistakes "due to deficiencies of the program while human players are continually guilty of very simple and obvious blunders."

This extended to errors of the psyche: computers couldn't suffer from a case of nerves or overconfidence, two deficits in human players that led to game-ending mistakes. A robot player could play emotionless, egoless chess: a clinical game in which each move was simply a new math problem.

But—and Shannon was emphatic about the "but"—"these must be balanced against the flexibility, imagination and inductive and learning capacities of the human mind." The great downfall of a chess-playing machine, Shannon thought, was that it couldn't learn on the fly, a capacity he believed was vital for victory at the elite levels. He cites Reuben Fine, an American chess master, on the misconceptions about top-ranked players and their approach to the game: "Very often people have the idea that masters foresee everything or nearly everything . . . that everything is mathematically calculated down to the smirk when the Queen's Rook Pawn queens one move ahead of the opponent's King's Knight's Pawn. All this is, of course, pure fantasy. The best course to follow is to note the major consequences for two moves, but try to work out forced variations as they go."

In mastering the probabilities of each conceivable position, then, a chess computer would not simply be acting as a superpowered grandmaster, but as a fundamentally different kind of player. Essentially, human and computer would be playing two different games while seated across the same board.

So Shannon cautioned against programming computers to behave too much like human beings: "It is not being suggested that we should design the strategy in our own image. Rather it should be matched to the capacities and weaknesses of the computer. The computer is strong in speed and accuracy and weak in analytical ability and recognition." Computers needed to be taken on their own merits and flaws, not as ersatz humans. What followed in the paper, and what Shannon would later popularize in a less technical article for *Scientific American,* was the range of strategies that could be programmed into a computer: a blueprint for turning a machine into a good, if not a great, player.

It is an admittedly broad survey: he studied each move's possible outcomes, considered game-theoretic approaches, outlined how a machine might go about evaluating moves, and concluded that a computer could be programmed to play a perfect game of chess, but that such an outcome would be wildly impractical. This was, in a way, a limitation of the technology of the time: if a contemporary computer's goal were to calculate all possible moves for itself and its opponent, it would not move its first pawn, Shannon calculated, for 10^{90} years.

Much as his information theory paper had done, Shannon's chess paper acted as a road map for an emerging field. Shannon would live to see the fruits of these labors; he would purchase machine after chess-playing machine, leading his exasperated wife to remark that "Claude went hog wild." But he took it one step further: Shannon's answer to Maelzel, one might say, came in the form of a machine he built himself. Completed in 1949, the machine was referred to as both Endgame and Caissac (after the fictional "patron goddess of chess," Caïssa). Shannon's machine could only handle six pieces and focused on the final moves in a chess game. Over 150 relay switches were used to calculate a move, processing power that allowed the machine to decide within a respectable ten to fifteen seconds.

The machine has largely been absent from accounts of Shannon's life. It is preserved at the MIT museum and in the memories of those closest to him. The box had the pattern of a chess board engraved on top of it; once the computer determined the correct move, a series of lights would indicate its choice to the user.

It was, by some accounts, the world's first chess-playing computer. It is also, perhaps more importantly, another illustration of Shannon's eagerness to build with his hands what he had dreamed up on paper.

For Shannon, both the chess paper and the chess machine addressed more enticingly ecumenical questions, as well. How should we think about "thinking machines"? Do machines *think* in the way we do? Do we want them to? What were an artificial brain's strengths and weaknesses? Shannon gave a measured answer, one that surely reflected the fact that he himself hadn't come to firm conclusions: "From a behavioristic point of view, the machine acts as though it were thinking. It has always been considered that skillful play requires the reasoning faculty. If we regard thinking as a property of external actions rather than internal method, the machine is surely thinking."

Shannon would, over time, grow more positive that artificial brains would surpass organic brains. Decades would pass before programmers would build a grand-master-level chess computer on the foundations that Shannon helped lay, but he was certain that such an outcome was inevitable. The thought that a machine could never exceed its creator was "just foolish logic, wrong and incorrect logic." He went on: "you can make a thing that is smarter than yourself. Smartness in this game is made partly of time and speed. I can build something which can operate much faster than my neurons." There was nothing more mysterious to it:

> I think man is a machine. No, I am not joking, I think man is a machine of a very complex sort, different from a computer, i.e., different in organization. But it could be easily reproduced—it has about ten billion nerve cells, i.e., 10^{10} neurons. And if you model each one of these with electronic equipment it will act like a human brain. If you take [Bobby] Fischer's head and make a model of that, it would play like Fischer.

25

Constructive Dissatisfaction

Shannon left little behind in the way of memoir, and the closest he ever came to autobiography was a talk he delivered in a Bell Labs auditorium in the same year that Theseus made its public debut. Fittingly, the talk revealed nothing of his background or his private life, but it was the kind of autobiography that mattered to him: a window into the workings of his brain. Ostensibly a lecture on "Creative Thinking," it turned out to be a tantalizingly brief tutorial on the appearance of the world from the eyes of a Shannon-level genius.

In one sense, the world seen through such eyes looks starkly unequal. "A very small percentage of the population produces the greatest proportion of the important ideas," Shannon began, gesturing toward a rough graph of the distribution of intelligence. "There are some people if you shoot one idea into the brain, you will get a half an idea out. There are other people who are beyond this point at which they produce two ideas for each idea sent in. Those are the people beyond the knee of the curve." He was not, he quickly added, claiming membership for himself in the mental aristocracy—he was talking about history's limited supply of Newtons and Einsteins. Of course, he was also lecturing a roomful of America's most gifted scientists on the prerequisites of genius, so one imagines that his humility only extended so far. In any case, once the prerequisites of talent and training had been satisfied, a third quality was still missing—something without which the world would have its full share of competent engineers but would lack even one real innovator.

It was here, naturally, that Shannon was at his fuzziest. It is a quality of "motivation . . . some kind of desire to find out the answer, the desire to find out what makes things tick." For Shannon, this was a requirement: "If you don't have that, you may have all the training and intelligence in the world, [but] you don't have the questions and you won't just find the answers." Yet he himself was unable to nail down its source. As he put it, "It is a matter of temperament probably; that is, a matter of probably early training, early childhood experiences." Finally, at a loss for exactly what to call it, he settled on curiosity. "I just won't go any deeper into it than that."

But then the great insights don't spring from curiosity alone, but from dissatisfaction—not the depressive kind of dissatisfaction (of which, he did not say, he had experienced his fair share), but rather a "constructive dissatisfaction," or "a slight irritation when things don't look quite right." It was, at least, a refreshingly unsentimental picture of genius: a genius is simply someone who is usefully irritated.

And finally: the genius must delight in finding solutions. It must have seemed to Shannon that though many around him were of equal intellect, not everyone derived equal joy from the application of intellect. For his part, "I get a big bang out of proving a theorem. If I've been trying to prove a mathematical theorem for a week or so and I finally get the solution, I get a big bang out of it. And I get a big kick out of seeing a clever way of doing some engineering problem, a clever design for a circuit which uses a very small amount of equipment and gets apparently a great deal of result out of it." For Shannon, there was no substitute for the "pleasure in seeing net results."

Presuming that one was blessed with the right blend of talent, training, curiosity, irritation, and joy, how would such a person go about solving an actual mathematical or design problem? Here Shannon was more concrete: he proposed six strategies, and the fluency with which he walked his audience through them—drawing P's for "problems" and S's for "solutions" on the chalkboard behind him for emphasis—suggests that these were all well-trodden paths in his mind.

You might, he said, start by simplifying: "Almost every problem that you come across is befuddled with all kinds of extraneous data of one sort or another; and if you can bring this problem down into the main issues, you can see more clearly what you're trying to do." Of course, simplification is an art form in itself: it requires a knack for excising everything from a problem *except* what makes it interesting, a nose for the distinction between accident and essence worthy of a scholastic philosopher. From the standpoint of Shannon's information theory, for instance, the difference between a radio and a gene is merely accidental, and yet the difference between a weighted and an unweighted coin carries essential weight.

Failing this difficult work of simplifying, or supplementing it, you might attempt step two: encircle your problem with existing answers to similar questions, and then deduce what it is that the answers have in common—in fact, if you're a true expert, "your mental matrix will be filled with P's and S's," a vocabulary of questions already answered. Call it ingenious incrementalism—or, as Shannon put it, "It seems to be much easier to make two small jumps than the one big jump in any kind of mental thinking."

If you cannot simplify or solve via similarities, try to restate the question: "Change the words. Change the viewpoint. . . . Break loose from certain mental blocks which are holding you in certain ways of looking at a problem." Avoid "ruts of mental thinking." In other words, don't become trapped by the sunk cost, the work you've already put in. There's a reason, after all, why "someone who is quite green to a problem" will sometimes solve it on their first attempt: they are unconstrained by the biases that build up over time.

Fourth, mathematicians have generally found that one of the most powerful ways of changing the viewpoint is through the "structural analysis of a problem"—that is, through breaking an overwhelming problem into small pieces. "Many proofs in mathematics have been actually found by extremely roundabout processes," Shannon pointed out. "A man starts to prove this theorem and he finds that he wanders all over the map. He starts off and proves a good many results which don't seem to be leading anywhere and then eventually ends up by the back door on the solution of the given problem." Fifth, problems that can't be analyzed might still be inverted. If you can't use your premises to prove your conclusion, just imagine that the conclusion is already true and see what happens—try proving the premises instead. Finally, once you've found your S, by one of these methods or by any other, take time to see how far it will stretch. The math that holds true on the smallest levels often, it turns out, holds true on the largest. "The typical mathematical theory is developed . . . to prove a very isolated, special result, [a] particular theorem. Someone always will come along and start generalizing it." So why not do it yourself?

In each of these methods, it's difficult to miss the echoes of Shannon's own work: the great simplification that turned computer relays into a shorthand for the language of logic, or the great generalization that identified the rules underlying every system of communication. Yet it is one thing to put these modes of thinking into words—and something else entirely to live inside them. Shannon seemed to recognize as much: "I think that good research workers apply these things unconsciously; that is, they do these things automatically." He went on to express his rationalist faith that any researcher would benefit from naming the tools, from making the unconscious conscious. But if it were really that simple, then why is it that "a very small percentage of the population produces the greatest proportion of the important ideas"? If there was any tension in the auditorium when he concluded—and invited the audience up to the front to examine a new gadget he'd been tinkering on—it was between Shannon the reluctant company man and Shannon the solitary wonder. The latter was as elusive as ever.

There is a famous paper on the philosophy of mind called "What Is It Like to Be a Bat?" The answer, roughly, is that we have no idea. What was it like to be Claude Shannon?

3

26

Professor Shannon

MIT made the first move: in 1956, the university invited one of its most famous alumni, Claude Shannon, to spend a semester back in Cambridge as a visiting professor. Returning to his graduate school haunts had something of a revivifying effect on Claude, as well as Betty. For one thing, the city of Cambridge was a bustle of activity compared to the comparatively sleepy New Jersey suburbs. Betty remembered it as an approximation of their Manhattan years, when going out to lunch meant stepping into the urban whirl.

Working in academia, too, had its charms. "There is an active structure of university life that tends to overcome monotony and boredom," wrote Shannon. "The new classes, the vacations, the various academic exercises add considerable variety to the life here." Reading those impersonal lines, one might miss the implication that Shannon himself had grown bored.

The work of teaching turned out to be a surprisingly pleasant change. A note from Shannon to a Bell Labs colleague gives a window into his new life as a professor:

> I am having a very enjoyable time here at MIT. The seminar is going very well but involves a good deal of work. I had at first hoped to have a rather cozy little group of about eight or ten

> advanced students, but the first day forty people showed up, including many faculty members from M.I.T., some from Harvard, a number of doctorate candidates, and quite a few engineers from Lincoln Laboratory. . . .
>
> I am giving 2 one and a half hour sessions each week, and the response from the class is exceptionally good. They are almost all following it at 100 percent. I also made a mistake in a fit of generosity when I first came here of agreeing to give quite a number of talks at colloquia, etc., and now that the days are beginning to roll around, I find myself pretty pressed for time. The people here are very interested in information theory, and there is a good deal of work going on both by the faculty and by graduate students specializing in that field.

The lecture audiences were as sharp as he might have hoped. "From the questions raised in the discussion period, I have a pretty favorable impression of the people attending," Shannon told another correspondent. "So far lecturing has not become a chore. In fact, I rather enjoy it, but I expect after a month or two, the novelty will wear off." It was, at least at the time, intellectually invigorating, not least because teaching was something Shannon had never done in any formal sense.

It was an opportunity, too, to glide across mathematics: freed of most professional obligations, Shannon was able to use each talk to dive deeply into a topic of personal interest. The "Seminar on Information Theory" in the spring term of 1956 served as a carousel for Shannon's passions. In a lecture titled "Reliable Machines from Unreliable Components," Shannon presented the following challenge: "In case men's lives depend upon the successful operation of a machine, it is difficult to decide on a satisfactorily low probability of failure, and in particular, it may not be adequate to have men's fates depend upon the successful operation of single components as good as they may be." What followed was an analysis of the error-correcting and fail-safe mechanisms that might resolve such a dilemma.

In another lecture, "The Portfolio Problem," Shannon pondered the implications for information theory of illicit gambling:

> The following analysis, due to John Kelly, was inspired by news reports of betting on whether or not the contestant on the TV program "$64,000 Question" would win. It seems that one enterprising gambler on the west coast, where the program broadcast is delayed three hours, was receiving tips by telephone before the local telecast took place. The question arose as to how well the gambler could do if the communication channel over which he received the tips was noisy.

And so on, like this. The lectures drew packed houses, including many members of the faculty who were busy with cutting-edge research of their own. Shannon and his musings, it seems, were draw enough to pull even the stars at MIT away from their work.

When an offer came for a full professorship and a permanent move to Massachusetts, it was hard to decline. If he accepted, Shannon would be named a Professor of Communication Sciences, and Professor of Mathematics, with permanent tenure, effective January 1, 1957, with a salary of $17,000 per year (about $143,000 in 2017). For all the pull of university life, Shannon struggled with his choice. Bell Labs had been his professional home for more than fifteen years. It had been the site of his most productive years as a researcher and thinker. It had afforded him unheard-of intellectual freedom and supported him in his most audacious pursuits. But Shannon was an outlier within the Labs culture; his antics were tolerated, but it was only a matter of time, Shannon suspected, before he would wear out his welcome. As he wrote to his supervisor, Hendrik Bode, "It always seemed to me that the freedom I took [at the Labs] was something of a special favor."

Bell Labs, understandably, didn't see it that way. They made a counteroffer, with a generous increase in Shannon's salary. But, in the end, it wasn't enough to sway him. His letter of resignation was a thoughtful weighing of industry against the academy. "There are certainly many points of superiority at Bell Labs," Shannon writes. "Perhaps most important among these is the freedom from teaching and other duties with a consequent increase in time available for research." Shannon acknowledged, too, that Bell Labs was offering him more money than MIT, "although the differential was not great in my case and, at any rate, I personally feel other issues are much more important."

Bell Labs' somewhat remote location in New Jersey was a complicating factor in its own right. "The essential seclusion and isolation of Bell Labs has both advantages and disadvantages. It eliminates a good many time-wasting visitors, but at the same time prevents many interesting contacts. Foreign visitors often spend a day at Bell Laboratories but spend six months at MIT. This gives opportunities for real interchange of ideas." Bell Labs matched and even exceeded MIT in the caliber of its thinking, Shannon allowed. But in the end, "the general freedom in academic life is, in my view, one of its most important features. The long vacations are exceedingly attractive, as is also the general feeling of freedom in hours of work." The two institutions are "roughly on par," which meant there was no one decisive factor pulling Shannon to MIT—only a certain restlessness on Shannon's part after spending more than a decade and a half in a single institution. "Having spent fifteen years at Bell Labs," Shannon writes, "I felt myself getting a little stale and unproductive and a change of scene and of colleagues is very stimulating."

Yet Shannon's associations with Bell Labs were, in the end, too strong for the Labs to simply sever all ties. Shannon was kept on the payroll. As the Labs' president, Bill Baker, later told Henry Pollak, "Shannon is one of the great people for which Bell Labs is known and going to be known. I will not take the chance of his ever being in poverty." Pollak would later joke that this was in keeping with the Bell Labs spirit: "There were two kinds of researchers at Bell Labs: those who are being paid for what they *used* to do, and those who are being paid for what they were *going* to do. Nobody was paid for what they were doing now." Perhaps in hopes of a return tour, Shannon's office was kept for him, his nameplate still gracing the closed door.

After accepting the MIT offer, the Shannons left for Cambridge via California—a year-long detour for a fellowship at Stanford's Center for Advanced Study in the Behavioral Sciences. Prestigious as the appointment was, the Shannons mainly treated it as an excuse to see the country. They made the leisurely drive through the West's national parks to California, and back, in a VW bus. Like many an East Coast professor before and since, Shannon marveled at Palo Alto and was said to have wondered aloud about how faculty there were able to finish any work in such luscious surroundings. Not long after, he recommended the same itinerary to a colleague: "You are going to God's country. All you need is a great white apron, a chef's hat, and a barbecue, and you'll be all set."

Before setting off for the West, though, Claude and Betty purchased a house at 5 Cambridge Street in Winchester, Massachusetts, a bedroom community eight miles north of MIT. Once their California year was complete, they returned to their new home. In Winchester, the Shannons were close enough to campus for a quick commute but far enough away to live an essentially private life. They were also living in a piece of history—an especially appropriate one in light of Shannon's background and interests.

Built in 1858, the house was constructed for Ellen Dwight, a great-granddaughter of a genius tinkerer of an earlier era, Thomas Jefferson. Originally seated on twelve acres, its design was inspired by Monticello. Encircled by "a three-sided verandah with segmental openings and chamfered posts," the house was a stately three stories at the crest of a "broad expanse of lawn reaching down to the wooded shore of Upper Mystic Lake." Toward the end of Shannon's life, it was added to the National Register of Historic Places, with the citation noting its "panoramic views of the lake and distant hills," as well as a sumptuous interior:

> The focal point of the plan is the first floor octagonal room. It contains a parquet floor said to be laid in a pattern identical to a floor at Monticello. The elaborate yellow marble fireplace surround has acanthus leaf, waterleaf and egg and dart moldings. Ceilings of the first story are approximately twelve feet in height; the ceilings are embellished with ornate plaster moldings around

> the perimeter. Windows of the lower story are full-length six-light windows, which when raised, provide egress to the verandah. The right parlor/library has a green marble fireplace.

The house would figure prominently in Shannon's public image. Nearly every story about him, from 1957 on, situated him at the house on the lake—usually in the two-story addition that the Shannons built as an all-purpose room for gadget storage and display, a space media profiles often dubbed the "toy room," but which his daughter Peggy and her two older brothers simply called "Dad's room."

The Shannons gave their home a name: Entropy House. Claude's status as a mathematical luminary would make it a pilgrimage site for students and colleagues, especially as his on-campus responsibilities dwindled toward nothing.

Even at MIT, Shannon bent his work around his hobbies and enthusiasms. "Although he continued to supervise students, he was not really a co-worker, in the normal sense of the term, as he always seemed to maintain a degree of distance from his fellow associates," wrote one fellow faculty member. With no particular academic ambitions, Shannon felt little pressure to publish academic papers. He grew a beard, began running every day, and stepped up his tinkering.

What resulted were some of Shannon's most creative and whimsical endeavors. There was the trumpet that shot fire out of its bell when played. The handmade unicycles, in every permutation: a unicycle with no seat; a unicycle with no pedals; a unicycle built for two. There was the eccentric unicycle: a unicycle with an off-center hub that caused the rider to move up and down while pedaling forward and added an extra degree of difficulty to Shannon's juggling. (The eccentric unicycle was the first of its kind. Ingenious though it might have been, it caused Shannon's assistant, Charlie Manning, to fear for his safety—and to applaud when he witnessed the first successful ride.) There was the chairlift that took surprised guests down from the house's porch to the edge of the lake. A machine that solved Rubik's cubes. Chess-playing machines. Handmade robots, big and small.

Shannon's mind, it seems, was finally free to bring its most outlandish ideas to mechanical life.

Looking back, Shannon summed it all up as happily pointless: "I've always pursued my interests without much regard to financial value or value to the world. I've spent lots of time on totally useless things." Tellingly, he made no distinction between his interests in information and his interests in unicycles; they were all moves in the same game.

Robert Gallager, decades later, would offer a comment that captures what many leading minds at the time likely thought of Shannon's private whimsies: "These were things that normal, outstanding scientists did not do!" Gallager was a Shannon disciple, and the remark was delivered with affection and only mock outrage, but it isn't too hard to imagine Shannon's more skeptical contemporaries wondering what the legend of Bell Labs could possibly have been thinking. His arrival at MIT, after all, carried with it great expectations. He had been awarded a named chair, tenure, and a position in two different departments, mathematics and engineering. "He was really lionized. He was going to be the luminary that led the electrical engineering department into the future of information theory," said Trenchard More, a former Shannon student.

Initially, it seems, Shannon's mere presence at MIT was powerful. It was a mark of distinction to have someone like him on the faculty, and he was useful in drawing energetic graduate students who might otherwise have gone elsewhere. Len Kleinrock, a graduate student of Shannon's from that era, recalled the impact of Shannon's arrival on his own decision making about graduate programs: "If I'm going to spend three or four years doing a PhD, I'm going to choose the best professor I can think of, and I want to do something with impact. The best professor, I knew, was Shannon."

Kleinrock wasn't the only one: graduate students in information theory were abuzz at the possibility of working with the inventor of their field. But the reality may have been somewhat less luminous. The few dissertation advisees he took on at MIT saw him infrequently. Asked to take on more students, he once responded, "I can't be an advisor. I can't give advice to anybody. I don't feel the right to advise." And it wasn't just Shannon's reticence: asking someone like Shannon for help induced anxiety in even the most capable. For Gallager, who began graduate study at MIT in the same year that Shannon joined the faculty, there was the small problem of asking a living legend to pencil him in:

> I was in such awe of him that I could hardly bring myself to speak to him! . . . He had very few doctoral students, and I think part of the reason was that, if you were at MIT with a colossal figure like Shannon around, you had to have a pretty big ego to ask someone like Shannon to supervise you!

Kleinrock put it perhaps more succinctly: "I always felt honored and a bit awkward that he'd be willing to work with me."

Somewhat inadvertently, Shannon played into this perception and kept himself at a remove from the normal comings and goings of academic life. He didn't join academic committees, jockey for status within his department, or even show up to his office with any regularity. What interaction he did have with his fellow faculty members usually took the form of dropping by unannounced at their lectures. One professor, Hermann Haus, remembered a lecture of his that Shannon attended. "I was just so impressed," Haus recalled, "he was very kind and asked leading questions. In fact, one of those questions led to an entire new chapter in a book I was writing."

Shannon lectured as well, dressed in coat and tie, as was expected of all MIT professors in those days, and occasionally flicking a piece of chalk in the air with one hand while answering student questions (and, impressively, never dropping the chalk). Once he was on faculty full-time, the lectures, it seems, received mixed reviews. Some students found them engaging and found Shannon to be just as good as advertised. "His classes were like a delicious meal! You'd go there, and the stuff he was giving you, it was clean, it was intuitive. It was good mathematics and it had impact," observed Kleinrock. For some, watching Shannon think out loud in a classroom setting would prove one of the defining moments of their academic lives.

Yet whatever challenges a genius might face explaining himself were evidently on full display in Shannon's lectures. The professor may have been enjoying himself, but some in his audience struggled to follow his train of thought. Dave Forney, then a student in the orbit around Shannon, observed that the quality of his talks depended almost entirely on the substance of the problem he had chosen to focus on that day. "For some problems, he had good results. For others, he made no progress beyond formulating the problem," said Forney, who added, "it was great for graduate students looking for a thesis topic."

In a way, even students who were fond of his lectures understood that they were, for Shannon, less a chance to impart specific information than to think out loud, to gather together MIT's finest and share some problem of personal interest. "He didn't teach that many classes," Kleinrock recalled. "I don't think he enjoyed it that much. He did it well, but I think he wanted to get the word out to this cadre of PhD students. Once he had taught them, he was happy working with them but not to continue teaching in that way to every generation." Or, as Gallager remembered:

> He was not the sort of person who would give a class and say "this was the essence of such and such." He would say, "Last night, I was looking at this and I came up with this interesting way of looking at it." He'd say it with a sly grin, and he'd come back with this absolutely beautiful thing.

This, then, was "Professor Shannon": too brilliant to be understood, or ignored. He was, by that point, more inspiration than instructor. Or, as one student put it, "We all revered Shannon like a god."

There were a lucky few students who managed to find a place at the deity's elbow. And for those who made it into Shannon's confidence, there were trips to the Winchester house and a standing invitation to bring him interesting problems. Kleinrock described his first interaction with Shannon: "He said, 'Why don't you come to my house next Saturday and visit with me?' and I said, 'Terrific.' You know, here I was, this lowly graduate student, I couldn't believe he had invited me to his house! . . . I remember telling my colleagues, 'I'm going to Shannon's house!' "

Shannon became a whetstone for others' ideas and intuitions. Rather than offer answers, he asked probing questions; instead of solutions, he gave approaches. As Larry Roberts, a graduate student of that time, remembered, "Shannon's favorite thing to do was to listen to what you had to say and then just say, 'What about . . .' and then follow with an approach you hadn't thought of. That's how he gave his advice." This was how Shannon preferred to teach: as a fellow traveler and problem solver, just as eager as his students to find a new route or a fresh approach to a standing puzzle.

Visits with Shannon generated their own folklore, and his suggestions during those sessions stayed with his students even decades after the fact. One anecdote, from Robert Gallager, captures both the power and subtlety of Shannon's approach to the work of instruction:

> I had what I thought was a really neat research idea, for a much better communication system than what other people were building, with all sorts of bells and whistles. I went in to talk to him about it and I explained the problems I was having trying to analyze it. And he looked at it, sort of puzzled, and said, "Well, do you really need this assumption?" And I said, well, I suppose we could look at the problem without that assumption. And we went on for a while. And then he said, again, "Do you need this other assumption?" And I saw immediately that that would simplify the problem, although it started looking a little impractical and a little like a toy problem. And he kept doing this, about five or six times. I don't think he saw immediately that that's how the problem should be solved; I think he was just groping his way along, except that he just had this instinct of which parts of the problem were fundamental and which were just details.
>
> At a certain point, I was getting upset, because I saw this neat research problem of mine had become almost trivial. But at a certain point, with all these pieces stripped out, we both saw how to solve it. And then we gradually put all these little

> assumptions back in and then, suddenly, we saw the solution to the whole problem. And that was just the way he worked. He would find the simplest example of something and then he would somehow sort out why that worked and why that was the right way of looking at it.

Other visitors, though, found themselves occasionally beaten to the punch by a mind that had mapped much of the terrain they were still only beginning to explore. Irwin Jacobs, an MIT student of that era and later the founder of Qualcomm, recalled: "People would go in, discuss a new idea, and how they were approaching it—and then he'd go over to one of his filing cabinets and pull out some unpublished paper that covered the material very well!"

Unlike traditional mid-century husbands and fathers, Shannon spent many of his days around the house. Of her many memories of her father, that would prove a distinctive one for Shannon's daughter, Peggy: "He did a lot of work at home so he would only go into the office to teach and to meet with graduate students, but if he didn't have to be there, he didn't spend much time at MIT. So my sense growing up was that he was around a lot. It was different from a lot of working people." Entropy House became his office; students dropped by, seeking feedback on projects but just as often looking to see what the Sage of Winchester had cooked up at his in-home laboratory. Even more conventional professors and old Bell Labs hands would make the trek to Winchester, and Shannon would walk them from room to room, all the while showing off his collection of contraptions and oddities. Guests were impressed by his collection of books, his two-story invention-room-cum-mechanic-shop, and the stunning array of gizmos and gadgets in the house.

It wasn't only Shannon's constant presence in the house, or the collection of electromechanical ephemera, that set him apart from other fathers. The Shannons were peculiar in the way that only a family headed by two mathematical minds might be. For instance, when it came time to decide who would handle the dishes after dinner, the Shannons turned to a game of chance: they wound up a robotic mouse, set it in the middle of their dining room table, and waited for the mouse to drop over one of the edges—and thus select that evening's dishwasher.

Then there were the spontaneous moments of math instruction. At a party hosted by the Shannons, young Peggy Shannon was in charge of the toothpicks. She was carrying a box of them on the house's verandah—and then dropped it by accident, spilling its contents onto the porch. Her father, standing nearby, paused, took stock of the mess, and then said, "Did you know, you can calculate pi with that?" He was referring to Buffon's Needle, a famous problem in geometric probability: it turns out that when you drop a series of needles (or toothpicks) on an evenly lined floor, the proportion of needles falling across a line can be used to estimate pi with surprising accuracy. Most important, Peggy remembered, her dad wasn't angry with her for the mess.

The Shannon household coalesced around the parents' passions: chess and music became family pastimes, and stock picking and tinkering were a part of everyday life. Shannon took his children to circus performances. *Alice in Wonderland,* the favorite of many a mathematician, was in the air; Shannon especially enjoyed quoting from "Jabberwocky." When it came to challenging math assignments, Peggy was regularly pointed in her father's direction, even though, as she admits, this was overkill; anyone in the household, including her two older brothers, could have helped. He was, by her account, a patient teacher, though he often went on tangents that betrayed his own inclinations. He complained about the educational fad of the "New Math" and would digress on concepts like imaginary numbers well past the point of helping his daughter finish her homework.

MIT, and its limited pressures on Shannon, also offered him the chance to step back from day-to-day work on information theory and observe the landscape of the still-coalescing digital world. Those years, said Thomas Kailath, who studied under Shannon, were something of "a golden age of information theory at MIT"—one in which Shannon played the role of godfather and network node, if no longer a central participant. Even if they had no direct contact with Shannon, a generation of new minds were brought into the field, with Shannon's work piquing their curiosity. As Anthony Ephremides, a later information theorist, put it, "The intellectual content of his approach was so appealing that many people who had inclinations to go in different directions said, 'Wow, I like this! This is a beautiful way of looking at this process about which I knew nothing, so let me find out more.'"

That more relaxed role might have seemed an indulgence, except that Shannon had, for all his humor and insouciance, been phenomenally productive by the time he left Bell Labs for MIT. Even with his aversion to writing things down, the famous attic stuffed with half-finished work, and countless hypotheses circulating in his mind—and even when one paper on the scale of his "Mathematical Theory of Communication" would have counted as a lifetime's accomplishment—Shannon still managed to publish hundreds of pages' worth of papers and memoranda, many of which opened new lines of inquiry in information theory. That he had also written seminal works in other fields—switching, cryptography, chess programming—and that he might have been a pathbreaking geneticist, had he cared to be, was extraordinary.

Yet Shannon had also come to accept that his own best days were behind him. "I believe that scientists get their best work done before they are fifty, or even earlier than that. I did most of my best work while I was young," Shannon said. This belief in an implicit age cap on mathematical genius was hardly unique to Shannon. As the mathematician G. H. Hardy famously wrote, "no mathematician should ever allow himself to forget that mathematics, more than any other art or science, is a young man's game."

While there have been notable exceptions to that rule, Shannon was convinced that he would not be one of them. His Bell Labs colleague Henry Pollak recalls visiting Shannon at home in Winchester to bring him up to date on a new development in communications science. "I started telling him about it, and for a brief time he got quite enthused about this. And then he said, 'Nuh-uh, I don't want to think. I don't want to think that much anymore.' It was the beginning of the end in his case, I think. He just—he turned himself off."

But if Shannon turned off the most rigorous part of his mind, he also freed himself to take a bird's-eye view of the emerging Information Age that his work had made possible. A crucial legacy of that work was the redirection of his colleagues' efforts. The old era had ended, one in which communications scientists were divided by medium, locked into fields of specialization whose gains did not shed any obvious light on one another.

"For everybody who built communication systems, before [Shannon], it was a matter of trying to find a way to send voice, trying to find a way to send data, like Morse code," recalled Gallager. "The thing that Claude said is that you don't have to worry about all those different things." Now their worries had a far more productive outlet: the coding, storage, and transmission of bits. "Once all the engineers were doing that, they start making this enormously rapid progress, start finding better and better ways of digitizing things and of storing and of communicating these very simple objects called binary digits, instead of these very complicated things like voice waveforms. If you look at it that way, Shannon is really responsible for the digital revolution."

And even though the revolution had begun to pass him by, Shannon's lectures at MIT and his talks around the country became a survey of the world to come. At a talk at the University of Pennsylvania in 1959, for instance, he said,

> I think that this present century in a sense will see a great upsurge and development of this whole information business . . . the business of collecting information and the business of transmitting it from one point to another, and perhaps most important of all, the business of processing it—using it to replace man at semi-rote operations at a factory . . . even the replacement of man in the things that we almost think of as creative, things like mathematics or translating languages.

If words like that seem self-evident and unremarkable to us today, it's worth remembering that Shannon was speaking more than a quarter century before the birth of the World Wide Web, and at a time when virtually all computers were still room-sized. To talk about "the information business" then was to talk about a world that was still more fantasy than fact.

So while it's a commonplace to say that Shannon's best thinking was over by 1948, that criticism might lead us to overlook a rich body of work, one marked by a playfulness of mind that was Shannon's lifelong calling card. Wish away the dilettante who spent the bulk of his later life on chess, machines, and juggling, and you'd also wish away the curious genius who invented information; it came, all of it, from the same place.

27

Inside Information

One of the great Shannon legends goes like this: in a fit of inspired mathematics, Shannon cracked a code for gaming the stock market. Huddled over old copies of the *Wall Street Journal,* Shannon put the full force of his mind to developing a series of algorithms that would make order of the market's chaos, giving him special insight into the tides of finance. It made him rich, and it might have made him the nation's leading investment guru, had he chosen to publicize his strategy.

As with most of the legends surrounding Shannon's life, it grew from a small grain of truth: in the 1960s and 1970s, Betty and Claude *did* play the markets obsessively. The process became a family affair, recalled Peggy Shannon:

> Much of the conversation around home would be about the stock market, because . . . much of my parents' focus was what was the market doing. They taught me to read the *Wall Street Journal* and the stocks very early. You'd come down and open the newspaper, and they'd have me read because my eyesight was better than theirs. And it was a way to engage the kids. . . . Then eventually they set up a small personal computer to carry the quotes during the day and then check again at the end of the

day, so there were computer printouts floating around the house with stock quotes on them.

By then the family had no need of the additional income from stock picking. Not only was there the combination of the MIT and Bell Labs pay, but Shannon had been on the ground floor of a number of technology companies. One former colleague, Bill Harrison, had encouraged Shannon to invest in his company, Harrison Laboratories, which was later acquired by Hewlett-Packard. A college friend of Shannon, Henry Singleton, put Shannon on the board of the company he created, Teledyne, which grew to become a multibillion-dollar conglomerate. As Shannon retold the story, he made the investment simply because "I had a good opinion of him." If there can be said to have been an old boys' club of Silicon Valley in its initial days, then Claude Shannon was a card-carrying member—and he benefited from all the privileges therein.

The club benefited from Shannon as well, in his roles as network node and informal consultant. For instance, when Teledyne received an acquisition offer from a speech recognition company, Shannon advised Singleton to turn it down. From his own experience at the Labs, he doubted that speech recognition would bear fruit anytime soon: the technology was in its early stages, and during his time at the Labs, he'd seen much time and energy fruitlessly sunk into it. The years of counsel paid off, for Singleton and for Shannon himself: his investment in Teledyne achieved an annual compound return of 27 percent over twenty-five years.

The stock market was, in some ways, the strangest of Shannon's late-life enthusiasms. One of the recurrent tropes of recollections from family and friends is Shannon's seeming indifference to money. By one telling, Shannon moved his life savings out of his checking account only when Betty insisted that he do so. A colleague recalled seeing a large uncashed check on Shannon's desk at MIT, which in time gave rise to another legend: that his office was overflowing with checks he was too absentminded to cash. In a way, Shannon's interest in money resembled his other passions. He was not out to accrue wealth for wealth's sake, nor did he have any burning desire to own the finer things in life. But money created markets and math puzzles, problems that could be analyzed and interpreted and played out. Shannon cared less about what money could buy than about the interesting games that money made possible.

The missing part of the story, as it happens, is Betty. The stock market intrigued her, and it was Betty, not Claude, who first drew the family into investing. She managed the family's finances—"I run the checkbook," she once told an interviewer. Peggy Shannon recalled that "their work in the stock market was completely a team effort. It is not the case that my father had these mathematical ideas about the stock market and then figured out how to put them to work to make money. . . . It was always a joint project." And it was made possible by the Shannons' shared tolerance for risk. As Peggy put it, "they were gamblers. They didn't shy away from making risky financial decisions."

Their kernel of interest in the market grew into a consuming hobby. The two of them, but Betty especially, began devouring books on trading, contemplating various market philosophies, and graphing possible scenarios for stocks. They studied many of history's most successful investors, including Bernard Baruch, Hetty Green, and Benjamin Graham. They read Adam Smith's *The Wealth of Nations* and studied Von Neumann and Oskar Morgenstern's work on game theory. Claude, unsurprisingly, contributed a device that was said to mirror how money flowed into and out of the market.

When Shannon offered to speak on the stock market at MIT, word of his talk forced him to relocate to the university's largest lecture hall, under its famous dome; even there, it was standing room only. Shannon proposed a theory that would allow an investor to profit from a stock whose value was *declining,* by making constant trades to take advantage of its price fluctuations. In answer to the very first question from the audience—Did he use this theory in his own investing?—he replied: "Nah, the commissions would kill you."

This talk—probably more than any particular feat of financial wizardry—is the main source of the legend of Shannon as stock-picking genius. Later Shannon seemed astounded by the attention his lecture received and unusually tickled when the subject came up in an interview:

> I even did some work on the theory of stocks and the stock market, which is among other papers that I have not published. Everybody wants to know what's in them! [*Shannon laughs.*] It's funny. I gave a talk at M.I.T. on this subject some twenty years ago and outlined the mathematics, but never published it, and to this day people ask about it. Just last year when we were over in Brighton more than one person came up to me and said, "Hi, heard you talked at M.I.T. about the stock market!" I was amazed that anybody would even have remembered it!

But for anyone searching for a grand unifying theory to explain the market's fluctuations, Shannon was quick to put such speculations to rest. He and his wife were, in his own words, "fundamentalists, not technicians." The Shannons had toyed with technical analysis, and they found it wanting. As Shannon himself put it, "I think that the technicians who work so much with price charts, with 'head and shoulders formulations' and 'plunging necklines' are working with what I would call a very noisy reproduction of the important data."

Complicated formulas mattered a great deal less, Shannon argued, than a company's "people and the product." He went on:

> A lot of people look at the stock price, when they should be looking at the basic company and its earnings. There are many problems concerned with the prediction of stochastic processes, for example the earnings of companies. . . . My general feeling is that it is easier to choose companies which are going to succeed, than to predict short term variations, things which last only weeks or months, which they worry about on *Wall Street Week*. There is a lot more randomness there and things happen which you cannot predict, which cause people to sell or buy a lot of stock.

From a lesser mathematical mind, this might have seemed like something of an evasion, but when Shannon used words like "stochastic processes," he spoke from deep experience with the underlying math. And it was his view that market timing and tricky mathematics were no match for a solid company with strong growth prospects and sound leadership.

So the Shannons sized up start-up founders in person whenever they could. They sampled products and prototypes. When they were mulling an investment in Kentucky Fried Chicken, as William Poundstone recounts, they bought several buckets' worth for a taste test with friends.

Beyond the research, there was another factor, one that Shannon was secure enough to readily acknowledge as key to his success. Asked if he was lucky in life, Shannon answered, "Far beyond any reasonable expectations." By his own admission, Shannon had been fortunate in his timing, and privileged in knowing certain company founders and securing early investments. The bulk of his wealth was concentrated in Teledyne, Motorola, and HP stock; after getting in on the ground floor, the smartest thing Shannon did was hold on. His daughter, Peggy, summed it up with a statement that could just as well have come from her father: her parents "used common sense and connections and had good luck."

If Shannon's work in the field of finance can be said to have left anything of lasting note, it's the memorable one-liners, many of which are among the best-known stories about him. "I make my money on the stock market. I don't make it by proving theorems," Shannon once famously told Robert Price. Asked what sort of information theory was best for investing, Shannon joked: "Inside information."

28

A Gadgeteer's Paradise

Many of Shannon's off-the-clock creations were whimsical—a machine that made sarcastic remarks, for instance, or the Roman numeral calculator. Still others showed a flair for the dramatic and dazzling: the trumpet that spit flames or the machine that solved Rubik's cubes. Still other devices he built anticipated real technological innovations by more than a generation. One in particular stands out, not just because it was so far ahead of its time, but because of just how close it came to landing Shannon in trouble with the law—and the mob.

Long before the Apple Watch or the Fitbit, what was arguably the world's first wearable computer was conceived by Ed Thorp, then a little-known graduate student in physics at the University of California, Los Angeles. Thorp was the rare physicist who felt at home with both Vegas bookies and bookish professors. He loved math, gambling, and the stock market, roughly in that order. The tables and the market he loved for the challenge: Could you create predictability out of seeming randomness? What could give one person an edge in games of chance? Thorp wasn't content just pondering these question; like Shannon, he set out to find and build answers.

In 1960, Thorp was a junior professor at MIT. He had been working on a theory for playing blackjack, the results of which he hoped to publish in the *Proceedings of the National Academy of Sciences*. Shannon was the only academy member in MIT's mathematics department, so Thorp sought him out. "The secretary warned me that Shannon was only going to be in for a few minutes, not to expect more, and that he didn't spend time on subjects (or people) that didn't interest him. Feeling awed and lucky, I arrived at Shannon's office to find a thinnish alert man of middle height and build, somewhat sharp featured," Thorp recalled.

Thorp had piqued Shannon's interest with the blackjack paper, to which Shannon recommended only a change of title, from "A Winning Strategy for Blackjack" to the more mundane "A Favorable Strategy for Twenty-One," the better to win over the academy's staid reviewers. The two shared a love of putting math in unfamiliar territory in search of chance insights. After Shannon "cross-examined" Thorp about his blackjack paper, he asked, "Are you working on anything else in the gambling area?"

Thorp confessed. "I decided to spill my other big secret and told him about roulette. Ideas about the project flew between us. Several exciting hours later, as the wintery sky turned dusky, we finally broke off with plans to meet again on roulette." As one writer put it, "Thorp had inadvertently set one of the century's great minds on yet another tangent."

Thorp was immediately invited to Shannon's house. The basement, Thorp remembered, was "a gadgeteer's paradise. . . . There were hundreds of mechanical and electrical categories, such as motors, transistors, switches, pulleys, gears, condensers, transformers, and on and on." Thorp was in awe: "Now I had met the ultimate gadgeteer."

It was in this tinkerer's laboratory that they set out to understand how roulette could be gamed, ordering "a regulation roulette wheel from Reno for $1,500," a strobe light, and a clock whose hand revolved once per second. Thorp was given inside access to Shannon in all his tinkering glory:

> Gadgets . . . were everywhere. He had a mechanical coin tosser which could be set to flip the coin through a set number of revolutions, producing a head or tail according to the setting. As

> a joke, he built a mechanical finger in the kitchen which was connected to the basement lab. A pull on the cable curled the finger in a summons. Claude also had a swing about 35 feet long attached to a huge tree, on a slope. We started the swing from uphill and the downhill end of the arc could be as much as 15 or 20 feet above the ground. . . . Claude's neighbors on the Mystic lake were occasionally astounded to see a figure "walking on the water." It was me using a pair of Claude's huge styrofoam "shoes" designed just for this.

And yet, Thorp wrote, what impressed him more than any of the gadgets was his host's uncanny ability to "see" a solution to a problem rather than to muscle it out with unending work. "Shannon seemed to think with 'ideas' more than with words or formulas. A new problem was like a sculptor's block of stone and Shannon's ideas chiseled away the obstacles until an approximate solution emerged like an image, which he proceeded to refine as desired with more ideas."

For eight months, the pair dove into the challenge of developing a device that would predict the final resting spot of a roulette ball. For the device to beat the house, Thorp and Shannon didn't have to predict the precise outcome every time: they just had to acquire any kind of slight edge over the odds. Over time, and with enough bets, even the smallest advantage would multiply into a meaningful return.

Picture a roulette wheel divided up into eight segments: by June 1961, Thorp and Shannon had a working version of a device that could determine which of those segments would end up holding the ball. As soon as they concluded that they had, in fact, found their edge, Shannon impressed upon Thorp the need for absolute secrecy. He invoked the work of social network theorists, who argued that two people chosen at random would be, at most, three degrees of separation from one another. In other words, the distance between Shannon, Thorp, and an enraged casino owner was slim.

The device that they created "was the size of a pack of cigarettes," operated by Thorp's and Shannon's big toes, "with microswitches in our shoes," and delivered gambling advice in the form of music. Thorp explained:

> One switch initialized the computer and the other timed the rotor and the ball. Once the rotor was timed, the computer transmitted a musical scale whose eight tones marked the rotor octants passing the reference mark. . . . We each heard the musical output through a tiny loudspeaker in one ear canal. We painted the wires connecting the computer and the speaker to match our skin and hair and affixed them with "spirit gum." The wires were the diameter of a hair to make them inconspicuous but even the hair thin steel wire we used was fragile.

They took it to the casinos, where Thorp and Shannon took turns placing bets. "The division of labor," Thorp said, "was that Claude stood by the wheel and timed, while I sat at the far end of the layout, unable to see the spinning ball well, and placed bets." Their wives served as lookouts, "checking to see whether the casino suspected anything and if we were inconspicuous." Even so, they had some close calls: "Once a lady next to me looked over in horror," Thorp recalled. "I left the table quickly and discovered the speaker peering from my ear canal like an alien insect."

Mishaps aside, Thorp was confident that the duo could run the tables. Claude, Betty, and Thorp's wife, Vivian, were less sure. Thorp would later concede that the others were probably on the right side of caution: the Nevada gaming industry was, notoriously, entangled with the mafia. Had Shannon and Thorp been caught, the odds were against two MIT professors talking their way out of it. The experiment was called off after its trial run, and the wearable computer was consigned to Shannon's growing heap of curiosities.

29

Peculiar Motions

"Do you mind if I hang you upside down by your legs?"

From any other professor, this question might have elicited concern. But from Claude Shannon, it was par for the course. Shannon had in mind an elaborate experiment: combining two forms of juggling—bounce and toss—by suspending a juggler by his feet.

Toss juggling is the airborne form of the art most familiar to us. Bounce juggling, on the other hand, consists in keeping objects in motion by hitting them against the ground in a pattern, a motion akin to the beating of hand drums. As generations of jugglers have discovered in the early stages of mastering their craft, bouncing items against the ground requires far less energy than tossing them in the air; in a bounce juggle, the balls arrive at the hand at the very top of their arc, the slowest speed they'll have during the entire sequence. But even though the bounce juggler has the advantage of catching a ball at the nadir of its speed, conventional juggling of the tossing-and-catching kind is a more fluid motion, one that comes more naturally to us and affords the juggler more control than the percussive effort of bounce juggling.

Shannon wondered: Was it possible to marry the physics of these two styles? Could you capture, in one motion, the fluidity of toss juggling and the efficiency of bounce juggling? In practical terms: If you were dangling by your feet, could you toss balls in the air, let gravity do the work of bringing the balls down to earth, and then catch them again? Both the inquiry and the method were vintage Shannon: whimsical, indifferent to practicalities, and originating in an activity that typical professors might have dubbed unserious, but which Shannon, a tenured member of the MIT faculty, found amusing enough to merit scholarly time and attention.

That's how Arthur Lewbel, an MIT student, found himself suspended by his feet in the middle of Shannon's living room. The balls went up . . . and dropped unceremoniously to the ground. "As a physical experiment, it was a complete failure," remembered Lewbel. Some physical limits even perfect math can't crack; and in this case, even the great Claude Shannon couldn't overcome the most obvious problem with the experiment's design: How well does anyone do anything upside down?

Lewbel had become accustomed to inquiries of the may-I-hang-you-upside-down variety. He was the founder of the MIT Juggling Club and first met Shannon when the famous information theorist casually dropped by the club's meeting unannounced. Shannon was there for the same reason that parents the world over find themselves in rooms not necessarily of their choosing: his daughter, Peggy, wanted him to go. She'd read about the club in the *Boston Globe*, and while it probably required minimal arm-twisting to get her unicycling, tinkering father to attend, the initial interest in the Juggling Club was Peggy's.

"He just showed up, and didn't tell anyone who he was. There were a bunch of jugglers standing outside practicing and he just walked up and said, 'Can I measure your juggling?' " remembered Lewbel. "That was the first thing he said to us, and it was something no one had ever asked us before." Lewbel and the other jugglers agreed to be measured, and Shannon and Lewbel developed a fast friendship.

Drop-ins from star faculty of Shannon's stature weren't unusual. As Lewbel tells it, "One nice thing about juggling at MIT is that you never know who will show up. For example, one day Doc Edgerton, inventor of the strobe light, stopped by the juggling club and asked if he could photograph some of us juggling under strobe lights." What was unusual was a return visit. But return Shannon did, repeatedly, even hosting the club at his Winchester home when they needed space for pizza-and-movie night. "The Juggling Club and the jugglers enchanted us," remembered Peggy Shannon.

Shannon had dabbled in juggling for decades. As a boy, he had imagined himself a fairground performer. At Bell Labs, the stories of his achievements in information theory were almost always accompanied by tales of his juggling while riding a unicycle in the Labs' narrow halls. At home in Winchester, there were ample objects stashed in the playroom to toss and catch. By that point, Shannon had developed his talents as an amateur juggler far past the point of mere amusement: he was said to be able to juggle four balls, which, as anyone who has tried to juggle knows, is a worthy achievement. Ronald Graham, a fellow mathematician-juggler, attributed some of his success to a trick borrowed from Galileo. "When Galileo wanted to slow gravity down, he just tilted a table" and let a ball roll from one end to the other, said Graham. "Imagine a big table, and then as you tilt the table, you get closer to 1 g." By sliding pucks up a tilted air hockey table, Shannon was able to study their patterns, and refine his juggling technique, in a kind of slow motion. The pucks' paths "were not parabolas, just pointy, and you could practice doing that."

Part of juggling's appeal to Shannon might have been the fact that it didn't come easily. For all his mathematical and mechanical gifts, "it was something he simply could not master, making it all the more tantalizing," wrote Jon Gertner. "Shannon would often lament that he had small hands, and thus had great difficulty making the jump from four balls to five—a demarcation, some might argue, between a good juggler and a great juggler." Here, at least, Shannon was destined to be merely good.

Juggling lacks the nobility of mathematical pastimes like chess or music. And yet the tradition of mathematician-jugglers is an ancient one. As best as we can tell, that tradition began in the tenth century

CE in an open-air market in Baghdad. It was there that Abu Sahl al-Quhi, later one of the great Muslim astronomers, got his start in life juggling. A few years later, Al-Quhi became a kind of court mathematician for the local emir, who, fascinated by planetary motion, built an observatory in the garden of his palace and put Al-Quhi in charge. The appointment bore some fine mathematical fruit: Al-Quhi invented an adjustable geometrical compass, likely the world's first, and led the revival among Muslim geometers of the study of the Greek thinkers Archimedes and Apollonius.

From juggling in a market to measuring the courses of planets: what they had in common, what drew Al-Quhi and so many of future number-crunching jugglers, was the patterns of parabolas and arcs, equations played out over open space. As Graham observed, "mathematics is often described as the science of patterns. Juggling can be thought of as the art of controlling patterns in time and space." So it's no surprise that generations of mathematicians could be found on university quads, tossing things in the air and catching them. Burkard Polster, author of *The Mathematics of Juggling,* writes, "next time you see some jugglers practicing in a park, ask them whether they like mathematics. Chances are, they do. . . . Most younger mathematicians, physicists, computer scientists, engineers, etc. will at least have given juggling three balls a go at some point in their lives."

So what drew Shannon to the study of juggling? "He liked peculiar motions. . . . I think what he liked about juggling was that it was a peculiar physical motion," noted Lewbel. It was in the early 1970s that these peculiarities finally tugged enough to provoke him to write a mathematical paper on the topic.

Juggling, observed Lewbel, "is complex enough to have interesting properties and simple enough to allow the modeling of these properties." But for all of its mathematical richness, when Shannon first began his work on the topic, he was starting from scratch: the field had no body of written work.

The first important scientific work on the topic was in the field of psychology. In 1903, Edgar James Swift published a paper in the

American Journal of Psychology that studied the time it took to learn how to juggle, as an examination of the most effective ways to teach neurosensory skills. The nature of juggling itself seems to have been something of an afterthought. The insight Swift was after wasn't "How does a juggler learn his craft?" so much as "How can a human being learn any craft?" Following in his footsteps, psychologists continued to use juggling as a research tool into the mid-twentieth century. But where psychologists had found juggling useful in their research, mathematicians had been reluctant to use a favorite pastime as a source of data and experiments. Until Shannon arrived on the scene, no papers had explored the math of juggling.

How could that be? How had millennia's worth of mathematicians tried their hands at juggling but published no mathematical results on the topic? In some ways, it's not hard to understand. Mathematics was then, as now, a fiercely competitive discipline, and while card games, puzzles, juggling, and other such entertainments may have been amusing mathematical hobbies, no serious, ambitious mathematician would have mistaken a circus routine for a topic deserving sustained research or publication. No one, that is, until Claude Shannon. Unmoved by material concerns, freed of the need to burnish his reputation, and driven by curiosity for curiosity's sake, he could throw himself headlong into the study of juggling without any of the misgivings his colleagues might feel about doing the same.

In the context of Shannon's other work, the juggling paper is unremarkable. It didn't inaugurate a new field of study, and it didn't bring him international acclaim. Shannon neither published it nor entirely finished it. Though Shannon was perhaps the first scientist to study juggling with mathematical rigor, the paper's striking feature isn't its originality or the quality of its mathematics, but rather what it reveals about its author's wide-ranging reading and research. If information theory, genetics, and switching proved the depth of Shannon's thinking, juggling displayed his dexterity. It is a testament, as well, to Shannon's belief that just about anything could be the object of serious mathematical analysis.

Shannon opens the paper with a dialogue from *Lord Valentine's Castle,* Robert Silverberg's science fiction novel set on the distant planet of Majipoor. It's a chronicle of the adventures of an itinerant juggler named Valentine, who, as it happens, is actually a king whose throne and title have been taken from him:

> "Do you think juggling's a mere trick?" the little man asked, sounding wounded. "An amusement for the gapers? A means of picking up a crown or two at a provincial carnival? It is all those things, yes, but first it is a way of life, friend, a creed, a species of worship."
>
> "And a kind of poetry," said Carabella.
>
> Sleet nodded. "Yes, that too. And a mathematics. It teaches calmness, control, balance, a sense of the placement of things and the underlying structure of motion. There is a silent music to it. Above all there is a discipline. Do I sound pretentious?"

For Shannon, it was important that people reading this paper "try not to forget the poetry, the comedy and the music of juggling for the Carabellas and Margaritas future and present." We can sense something of Shannon's self-consciousness in the next sentence, when he interrupts this train of thought and cribs Sleet to ask the reader, "Does this sound pretentious?"

If it did, Shannon knew it. It might explain why his next paragraphs seek to soften the lofty-sounding beginnings of the paper by grounding the reader in the history of juggling. In the span of roughly two pages, he travels over 4,000 years and covers a considerable range of popular and cultural nods to juggling. The paper's historical tour opens in early Egypt, circa 1900 BCE, in tombs with juggling scenes etched into the walls, four women each tossing three balls apiece. From there it's off to the Polynesian island of Tonga, with the sailor-adventurer Captain James Cook and scientist Georg Forster. The year was 1774, and Forster observed, in *A Voyage Round the World,* that the Tongans had a flair for keeping multiple objects suspended in the air in sequence. Shannon quotes Forster's observation of one girl who, "lively and easy in all her actions, played with five gourds, of the size of small apples, perfectly globular. She threw them up into the air one after another continually, and never failed to catch them all with great dexterity, at least for a quarter of an hour."

From there it was back to dry land and 400 BCE, to Xenophon's *The Banquet* and an audience with Socrates, who, upon seeing a young woman juggle twelve hoops in the air, is moved to observe, "This girl's feat, gentlemen, is only one of many proofs that woman's nature is really not a whit inferior to man's, except in its lack of judgment and physical strength. So if any one of you has a wife, let him confidently set about teaching her whatever he would like to have her know." For Shannon, Socrates's comment is interesting on two levels. For one thing, if the girl in this scene did in fact juggle twelve hoops, she would hold the world record for the most objects juggled at a single time. On this fact Shannon is willing to give Xenophon and Socrates the benefit of the doubt: "Who could ask for better witnesses than the great philosopher Socrates and the famous historian Xenophon? Surely they could both count to twelve and were careful observers."

But that's Shannon's only concession; Socrates's chauvinism doesn't sit well. Mustering as much antagonism as we might imagine Shannon capable of, he dismisses Socrates's blinkered view of a woman's capacities. "It is amusing to note how Socrates, departing from his famous method of teaching by question, makes a definite statement and immediately suffers from foot-in-mouth disease. Had he but put his period nine words before the end he could have been the prescient prophet of the women's equality movement." Later in the paper, Shannon makes the case for female jugglers more explicit. Two he singles out for special mention: Lottie Brunn, the "world's fastest female juggler" and a fixture on the 1920s European theater circuit; and Trixie Firschke, the "first lady of juggling," a German child star born to a Budapest circus family.

So, beginning in ancient Egypt and passing through the medieval minstrel's hybrid of "juggling, magic, and comedy," Shannon ends in the world of twentieth-century variety shows. Their leading lights—including W. C. Fields—inspired a generation of girls and boys, including the young Claude Shannon, to terrify their parents with talk of running away to join the circus.

The history lesson concluded, it was on to a more serious inquiry: how to understand the psyche of a juggler and the practice of juggling? Specifically, how does one make sense of a practice that requires precision but is also the stuff of comedy? A gymnast's fault elicits pity, a kind of disappointment shared by performer and audience; a juggler, failing to catch a club, is just as likely to be met with laughter. How do jugglers deal with this?

"Jugglers are surely the most vulnerable of all entertainers," Shannon writes, walking up again to the edge of autobiography. Indeed, most serious jugglers are forced to develop a range of mind games and public feints to deal with the anguish of "the missed catch or the dropped club." Their coping strategies vary with skill level: lesser jugglers paper over their failures with comedy and props; the experts make their failures appear as intentional as their successes.

But this vulnerability is precisely why, Shannon notes, juggling's ranks can be roughly divided into two camps: performance jugglers and technical jugglers. Technicians are in a numbers game, an arms race of objects juggled. The more objects in the air, the bigger the bragging rights. Shannon references one of the world's great technicians, Enrico Rastelli, about whom *Vanity Fair* would say in eulogy: "In his twenty years' devotion to his craft this son of Italy elevated it, for probably the first time, to what was unmistakably an art." Rastelli, Shannon noted, was able to keep ten balls in the air simultaneously. Shannon also remarks that Rastelli "could do a one-armed handstand while juggling three balls in the other hand and rotating a cylinder with his feet."

Rastelli and his breed of technical juggler held the most interest for Shannon, and mathematicians since. Call it seriousness of purpose, or the possibility of organizing by numbers and implicit formulas the quest to manage an ever-increasing number of objects. For the mathematician, performance juggling, enjoyable though it might be to watch, possesses none of these qualities. The joy of the crowd, the thrill of the motion, the comedy of the effort—all of these are amusing but ultimately uninteresting to a mind trained in math. The paper's journey begins here: in the challenge of increasing the number of objects juggled while still maintaining precision, the intersection of mathematics and movement.

It shouldn't come as any particular surprise that Shannon, whose love of juggling was surpassed only by his love of music, opens the mathematical section of the paper with a reference to jazz—in particular, to the drummer Gene Krupa, who said that "the cross rhythm of 3 against 2 is one of the most seductive known." For Shannon, the pattern of three against two is a useful analogue for an introduction to the mathematics of juggling. It's the pattern by which most people first learn to juggle: three balls in two hands.

Pick apart the motions of a juggler and what emerges are a series of predictable parabolas. One ball tossed in the air produces one arc; multiple balls, multiple arcs. All that's left is to combine them into a consistent pattern, set to a rhythm. This was how Shannon approached the problem of juggling—not only as an exercise in coordination, but as an algebraic formula. His juggling theorem stated the following:

$(F + D) H = (V + D) N$
F = how long a ball stays in the air
D = how long a ball is held in a hand
H = number of hands
V = how long a hand is empty
N = number of balls being juggled

Shannon's theorem tracks time continuously. As Lewbel put it, "The way the juggler achieves the rhythm in Shannon's theorem is by trading off time in a continuous way; the more time one ball spends in the air relative to the time it spends in your hand, the more time you have to deal with the other balls, and so the more balls you can juggle. Shannon's theorem makes this trade off in times precise." (He also pointed out the irony, given the rest of Shannon's digital innovations, that the juggling theorem's measurement of continuous time makes it analog.) Each side of the equation tracks a different part of the act of juggling: the left side tracks the pattern of the balls, and the right side tracks the pattern of the hands. Because, as Lewbel put it, "the amount of time balls spend being juggled is the same as the time the hands spend juggling them," the equality is maintained.

Shannon's work on juggling might have ended here; already he had lent the study of juggling considerable legitimacy, and given a generation of mathematician-jugglers the ability to combine their two passions without fear of embarrassment. But a paper was, in this case, insufficient. In 1983, as he had so often before, Shannon brought the work from the world of theory into the realm of mechanics: he set out to build his own juggling robot.

"It all started when Betty brought home a little four-inch clown, doing a five ball shower, from the cake decorating store ($1.98). I was both amused and bemused—amused as a long-time amateur juggler who even as a boy wished to run away and join the circus, but bemused by the unlikely shower pattern and the plastic connections between the balls," Shannon wrote.

The cake-store clown only appeared to juggle—but Shannon's robot actually did. Assembled from his Erector Set, the finished product was able to handle three balls. The balls bounce off a tom-tom drum, and the robot moves its paddle arms in a rocking motion, "each side making a catch when it rocks down and a toss when it rocks up." Though Shannon never completed the bounce-juggling robot's counterpart—a robot that could authentically toss juggle—he still built his own set of clowns that put on a convincing imitation. And there was one way, he noted with pride, in which they outclassed any human: "The greatest numbers jugglers of all time cannot sustain their record patterns for more than a few minutes, but my little clowns juggle all night and never drop a prop!"

30

Kyoto

For decades, honors and recognitions arrived for Shannon from around the world. The world's top universities conferred honorary degrees. Societies of all sizes bestowed certificates, commendations, and gold medals.

The boy from Gaylord was mostly amused by all the attention. As Betty Shannon later put it, "He was a very modest guy. He got a lot of awards but they never went to his head and he never talked about them." Shannon put it this way:

> I don't think I was ever motivated by the notion of winning prizes, although I have a couple of dozen of them in the other room. I was more motivated by curiosity. Never by the desire for financial gain. I just wondered how things were put together. Or what laws or rules govern a situation, or if there are theorems about what one can't or can do. Mainly because I wanted to know myself.

Shannon's indifference was on display for all to see: he had accumulated so many honorary degrees that he hung the doctoral hoods from a device that resembled a rotating tie rack (which, naturally, he had built with his own hands). Whether the awarding institutions would have found that treatment fitting or insulting, it speaks to the lightness with which Shannon took the work of being lauded.

Peggy's account of those years gives the impression of parents trying to keep their home life normal in the face of their Claude's fame in the mathematics world. "Then," Peggy remembered, "the calls would come about the honorary degrees," cracking the thin veneer.

Try as they might to downplay and laugh off many of his achievements, some honors made it clear even to a child that Claude was someone important—and that, as unassuming as he was, his work constituted something significant in the world. On the day before Christmas 1966, it was announced that President Lyndon B. Johnson would present Claude Shannon with the National Medal of Science, in honor of "brilliant contributions to the mathematical theories of communications and information processing."

On February 6, 1967, the Shannons joined the assembled guests in the East Room of the White House, where President Johnson dedicated his remarks to "eleven men whose lifelong purpose has been to explore the great ocean of truth. Their achievements—and the work of other scientists—have lengthened man's life, have eased his days, and have enriched our treasury of wisdom." It was a proud day for the Shannon family, all of whom were in attendance. Peggy recalled that she argued with her mother over which dress to wear, but also remembered, as many a White House guest does, the feeling of significance conveyed simply by setting foot in the building. She remarked, channeling some of her father's self-effacement, "I was seven, so I had the eyes of a seven-year-old, and it was just a pretty cool thing."

LBJ gave the family the full Johnson treatment just after presenting the award, and Vice President Hubert Humphrey's loud laugh spooked young Peggy into hiding behind her mother's leg.

Among Shannon's most cherished awards and honors were the ones that made him laugh. There was a miniature Greek temple, bearing the inscription "MASSACHVSETTS INSTITVTE OF JVGGLOLOGY" and featuring a clown juggling tiny replicas of Shannon's honorary diplomas. And, from the close of his fellowship at Stanford, a very formal, university-sanctioned certificate—the bottom of which had been graffitied with the signatures of all the other fellows, in as large and boisterous a script as the space would allow. Shannon even found a way to wring comedy from accepting awards. Upon being invited to join the American Philosophical Society, he was sent a certificate that was an obvious facsimile of calligraphy. Tickled, Shannon hired an *actual* calligrapher to write a long reply accepting membership into the society.

Not even the Oxbridge style of high academic decorum could dent his flippancy. When he was awarded a visiting fellowship to All Souls College in Oxford in 1978, he had the opportunity of reuniting with John Pierce and Barney Oliver for the university's Trinity term. The trio, along with their fellow Bell Labs alumnus and reunion organizer Rudi Kompfner, were expected to give a series of lectures on their topics of research and interest—artificial intelligence, information theory, and so on. Notes passed between Kompfner and Pierce testify to their concern about Shannon's willingness to deliver the lectures: "to get something out of Claude may be the problem," Kompfner wrote to Pierce.

Claude was pondering a serious problem, though—or at least, a serious problem for him. What emerged from the Oxford stint was one of the more curious papers of Shannon's career. Frustrated by having to drive on the left side of the road, Shannon engineered a custom-built solution. "The Fourth-Dimensional Twist, or a Modest Proposal in Aid of the American Driver in England" opens with a tale of the woes of the American driver abroad:

> An American driving in England is confronted with a wild and dangerous world. . . . With our long-ingrained driving habits the world seemed totally mad. Cars, bicycles and pedestrians would dart out from nowhere and we would always be looking in the wrong direction. The car was usually filled with curses from the men and with screams and hysterical laughter from the women as we careened from one narrow escape to another. The passengers were given to sudden involuntary motions—shielding the face or slamming on non-existent brakes. The turn indicator and

> windshield wiper controls were also reversed from American practice and we found ourselves signaling turns with the windshield wiper—fast for a right turn, slow for a left. The whole driving situation was not particularly improved by the narrowness of English streets and the high speed of English drivers. Nor was our inner security increased by the predilection of the English for building stone walls immediately adjacent to the roads.

Shannon proposed an idea that even he admitted sounded "grandiose and utterly impractical—the idle dream of a mathematician." His solution was to create a fourth dimension, one that reversed perceptions of right and left:

> How will we do this? In a word, with mirrors. If you hold your right hand in front of a mirror, the image appears as a left hand. If you view it in a second mirror, after two reflections it appears now as a right hand, and after three reflections again as a left hand, and so on. Our general plan is to encompass our American driver with mirror systems which reflect his view of England an odd number of times. Thus he sees the world about him not as it is but as it would be after a 180° fourth-dimensional rotation.

Finally, a series of adjustments to the steering system would translate the American driver's motions into British English: turning the wheel left would make the car go right, and vice versa, *et voilà*.

Complete with drawings, figures, and schematics, the paper was, of course, written with tongue firmly in cheek. But it remains the most memorable record of Shannon's time at Oxford. At more than 2,100 words, it was not simply a throwaway idea—it shows Shannon's willingness to spend hours fleshing out the implications of a joke, as well as his imperturbable indifference to the honors that came his way. And it speaks, perhaps, to the minor anxieties of a world traveler who mainly found travel something to be tolerated—who would just as soon have brought his home with him, even if only as an optical illusion.

By the time the awards-related junkets began in earnest, the Shannons had three children, so each trip became a chance for the family to travel the world together. His daughter, Peggy, recalled, "he'd get an award from Israel and the whole family went on a six- or seven-week trip in the middle of the school year. We went to Israel and then we went to Egypt, Turkey, and England. . . . So I was pulled out of school for six weeks or something like that in order to do that."

Shannon himself had mixed feelings about all of the travel. He was a homebody and an introvert, and more important, a less-than-adventurous eater. His tastes ran to home-cooked meat and potatoes, and the problem of finding the closest foreign equivalent caused him no small amount of anxiety. Peggy remembered that the Shannons rarely ate out, even in Massachusetts—so the prospect of couscous in Israel or raw fish in Japan was particularly fearsome to her father.

Nor did it help that public speaking increasingly terrified him, especially, it seems, as he grew further removed from the work that made his name. The confident, if occasionally scattershot, MIT lecturer had developed a crippling stage fright—driven less, it seems, by fear of the spotlight than by fear of having run out of interesting, intellectually rigorous subjects to talk about. For Shannon there would be little of the aging luminary's pontifications and platitudes; the standard he generally set himself was hard mathematics or nothing.

Even sympathetic audiences and eponymous venues frightened him. In 1973, for instance, Shannon was invited to give the first Claude Shannon lecture in Ashkelon, Israel, for the Institute of Electrical and Electronics Engineers Information Theory Society. "I have never seen such stage fright," the mathematician Elwyn Berlekamp recalled. "It never would have occurred to me that anyone in front of friends could be so scared." Shannon required extensive nerve calming in the wings and would only take the stage accompanied by a friend. Another attendee remembered, "He just felt that people were going to expect so much of him in this talk, and he was afraid that he didn't have anything significant to say. Needless to say, he gave a fantastic talk, but in my mind . . . it showed me what a modest man he was."

In response to another invitation from a friend, Shannon anticipated that he'd be asked to speak—and tried a preemptive strike: "Since our retirement, Betty doesn't do windows and I don't give talks." Still, for all his anxieties about appearing in public, Shannon indulged in the trips and accepted the accolades, if only because Betty so enjoyed the chance to see the world.

In part, the invitations and recognitions kept pouring in because the technological developments of the 1970s had awakened the world to information theory's importance. In the immediate aftermath of Shannon's "Mathematical Theory of Communication," said an MIT student of that era, Tom Kailath, "we always thought that information theory would never see practical implementation. In the old days, people studied Latin and Greek just as training for the mind." In the same way, young engineers in the 1950s and '60s mainly saw Shannon's theory as "good training."

And yet an increasingly digital world had begun to assimilate the codes whose existence Shannon had first identified. On September 5, 1977, the Voyager 1 probe set out for Jupiter and Saturn, fortified against error by one of those codes and capable of transmitting images of the gas giants across some 746 million miles of vacuum. In the same year, a pair of Israeli researchers, Jacob Ziv and Abraham Lempel, devised a data compression algorithm, built on Shannon's coding work, that served as one of the critical backbones of later Internet and cellular communications systems. That Ziv had been a graduate student at MIT at the same time as Shannon was a faculty member was, by his own later acknowledgment, critical to firing his interest in the field.

Even as the scale of Shannon's accomplishment became increasingly clear, "he didn't like to boast," remembered Arthur Lewbel.

> But every now and then . . . I remember one time I was at his house, and he was showing me the program for an information theory conference. He just picked it up, showed it to me, and he pointed to the sessions. And the session was named Shannon Theory 1, and another session he pointed to was named Shannon Theory 2, and the sessions went up to Shannon Theory 5.

Naturally, talk of a Nobel Prize followed Shannon for much of his career. In 1959, he was nominated for the Nobel in physics, alongside Norbert Wiener. Instead, physicists Emilio Gino Segrè and Owen Chamberlain garnered that year's award for discovering the antiproton. The nomination for Shannon and Wiener was regarded as a bit of a long shot, but the mere fact of a nomination reveals what Shannon's contemporaries thought of him. The problem with awarding Shannon the Nobel was, in part, structural: mathematics lacked a dedicated Nobel category of its own, a fact that had always been something of a chip on the math world's shoulder. Shannon himself said as much: "You know, there's no Nobel in mathematics, although I think there should be." The mathematicians who had won, including John Nash and Max Born, had won in fields like economics or physics; Bertrand Russell won in literature. Shannon's work had cut across several disciplines, but it was difficult to shoehorn into any Nobel field; the prize would not be in his future.

In 1985, though, the Shannons received a call, not from Stockholm, but from Kyoto. Claude had been selected as the first-ever winner of the Kyoto Prize in Basic Science, an award endowed by the billionaire Kazuo Inamori. Inamori was a Japanese applied chemist who founded the multinational Kyocera and later rescued Japan Airlines from bankruptcy. He was an engineer by training, a Zen Buddhist by choice, and a business turnaround artist by reputation. He was a student of management philosophy, which, along with his Buddhism, might explain why the Kyoto Prize's founding letter reads like a curious mix of a spiritual text and a shareholder update:

> After a quarter of a century of relentless and painstaking effort, Kyocera's annual sales have, by the grace of God, grown to 230 billion yen, with pre-tax profits of 53 billion yen. . . . I have decided on this occasion to create the Kyoto Prize. . . .
>
> Those worthy of the Kyoto Prize will be people who have, as have we at Kyocera, worked humbly and devotedly, sparing no effort to seek perfection in their chosen professions. They will be individuals who are sensitive to their own human fallibility and who thereby hold a deeply rooted reverence for excellence. . . .

> The future of humanity can be assured only through a balance of scientific progress and spiritual depth. Though today's technology-based civilization is advancing rapidly, there is a deplorable lag in inquiry into our spiritual nature. I believe that the world is composed of mutual dichotomies—pluses and minuses, such as the yin and the yang or darkness and light. Only through the awareness and nourishment of both sides of these dualisms can we achieve a complete and stable equilibrium. . . . It is my sincere hope that the Kyoto Prize may serve to encourage the cultivation of both our scientific and spiritual sides.

In time, the Kyoto would acquire a measure of prestige, in part by conspicuously styling itself as a competitor to the Nobel. Press releases announcing the prize winners would begin: "The Kyoto Prize, alongside the Nobel Prize one of the world's highest honors for the lifetime achievement of outstanding personalities in the fields of culture and science, is being awarded this year to . . ." It even managed to anticipate the Nobel on a number of occasions, honoring scientists who, years later, strained to avoid repeating themselves at their laureate lectures in Stockholm.

In presentation, too, the Kyoto ceremony would reach for Nobel-like flourish and flair, with the Japanese imperial family lending itself to the proceedings. And perhaps betraying its founder's sense for untapped business opportunities, the Kyoto's categories were broad enough to accommodate the fields the Nobel excluded—including mathematics and engineering. The Nobel may have had an eighty-four-year head start, but the Kyoto intended to give it a run for its money.

The Kyoto was a significant triumph for Shannon and represented, in many ways, his career's crowning recognition. Shannon, as usual, was nervous about the trip, and especially apprehensive about Japanese food. But he was joined by both Betty and his sister, Catherine, who still shared the family passion for math: she was a professor of mathematics at Murray State University in Kentucky. Accompanied as he would be, in Peggy's words, by "two strong women," he agreed to travel to Japan to accept the award.

Shannon's Kyoto Prize had a lasting benefit that outlived the award proceedings: he was required to deliver a laureate lecture, one of his last and longest public statements, "Development of Communication and Computing, and My Hobby." Shannon began the lecture by discussing history itself—or rather, the problem of how history was taught in his home country:

> I don't know how history is taught here in Japan, but in the United States in my college days, most of the time was spent on the study of political leaders and wars—Caesars, Napoleons and Hitlers. I think this is totally wrong. The important people and events of history are the thinkers and innovators, the Darwins, Newtons and Beethovens whose work continues to grow in influence in a positive fashion.

One category of innovation he singled out for special mention: the discoveries of science "are wonderful achievements in themselves, but would not affect the life of the common man without the intermediate efforts of engineers and inventors—people like Edison, Bell and Marconi." Shannon marveled at the progress of the twentieth century, before which "people lived much as they had centuries before, a largely agrarian society with little mobility or distant communication." He cited the spinning jenny, Watt's steam engine, the telegraph, the electric light, the radio, and the automobile—all less than two centuries old and each one transformative. That human life had been so utterly reshaped over a handful of life spans was largely, he believed, the work of engineers.

Though he was rarely given to public self-reflection, Shannon recalled the day when, as a young engineering student, he was asked to purchase a slide rule, a log-log-duplex, "the biggest they had." Looking back, he remarked on how quaint the device seemed now. He had with him on stage a handheld transistor computer made in Japan, and it "does everything my log-log-duplex did, and much more—and out to ten decimal places instead of three."

Between the slide rule and the handheld computer—between the room-sized differential analyzer and the Apple II sitting on his desk at home—Shannon's career had spanned a computing revolution. At points, "the intellectual progress in computers . . .was so rapid that they were obsolete even before they were finished."

In a room of Japanese royalty and distinguished guests, Shannon was giving an all-too-brief survey course of computing history, down to the point at which Shannon himself entered the story. It was the summation of a lifetime's work on machines that could communicate, think, reason, and act, and on the theoretical architecture that made all of it possible. But computing was not only a central thread of his life's work. As the lecture's title suggested, it was also, always, his hobby—or, as he translated the word for his audience, his *shumi*. "Building devices like chess-playing machines and juggling robots, even as a '*shumi*,' might seem a ridiculous waste of time and money," Shannon admitted. "But I think the history of science has shown that valuable consequences often proliferate from simple curiosity."

What might proliferate from such curiosities as Endgame and Theseus?

> I have great hopes in this direction for machines that will rival or even surpass the human brain. This area, known as artificial intelligence, has been developing for some thirty or forty years. It is now taking on commercial importance. For example, within a mile of MIT, there are seven different corporations devoted to research in this area, some working on parallel processing. It is difficult to predict the future, but it is my feeling that by 2001 AD we will have machines which can walk as well, see as well, and think as well as we do.

But even before this convergence of human and machine intelligence had come to pass, machines were still a rich source of analogies for understanding the subtleties of our own minds:

> Incidentally, a communication system is not unlike what is happening right here. I am the source and you are the receiver. The

translator is the transmitter who is applying a complicated operation to my American message to make it suitable for Japanese ears. This transformation is difficult enough with straight factual material, but becomes vastly more difficult with jokes and double entendres. I could not resist the temptation to include a number of these to put the translator on his mettle. Indeed, I am planning to take a tape of his translation to a second translator, and have it translated back into English.

We information theorists get a lot of laughs this way.

31

The Illness

She is leaving him, not all at once, which would be painful enough, but in a wrenching succession of separations. One moment she is here, and then she is gone again, and each journey takes her a little farther from his reach. He cannot follow her, and he wonders where she goes when she leaves.

—**Debra Dean**

To friends, the first noticeable signs of the disease came in the early 1980s. Initially, they noticed that he struggled to answer familiar questions. Then came the brief lapses in memory. In the earliest stages, some friends thought nothing of it. Shannon's achievements, after all, had been triumphs of intuition and analysis, not memory or recall. As Robert Gallager put it, "Claude was never a person who depended a great deal on memory, because one of the things that made him brilliant was his ability to draw such wonderful conclusions from very, very simple models. What that meant was that, if he was failing a little bit, you wouldn't notice it." That he was beginning to forget things was, to many of those closest to him, simply a sign that he was succumbing to the normal hazards of age.

Soon, though, he was forgetting the way home from the grocery store and unable to remember phone numbers, names, faces. His hand began to quiver when he wrote. Peggy Shannon remembered one day when the family was hosting the Juggling Club. She was sitting on the floor, her father in a chair nearby. He looked at her, paused, and asked, "Do you juggle?"

"I was just floored," she recalled. "Is it that he doesn't know who I am or that he doesn't remember that I juggle? Either one of them would be totally devastating."

By then, the marked change in Claude was undeniable. "In 1983, he went to a doctor," Betty said, "and the decision was that he probably was in the very, very early stages of Alzheimer's."

The Shannons began to be more deliberate about which travel invitations they accepted and which they rejected. At a 1986 event at the University of Michigan, Shannon was "very quiet," noted organizer David Neuhoff. "I had the feeling at that time that he was already suffering from the affliction of Alzheimer's. Betty did most of the talking." Decisions about how much travel to attempt or how much information to reveal about Shannon's illness fell to Betty, who desired to keep a cone of privacy around the family. "They felt like they'd earned the right to be private," Peggy Shannon recalled.

The family worked to keep his mind engaged, but the disease took its toll. Shannon lost a great deal of cognitive function very quickly, and the challenges of caring for someone with Alzheimer's put a heavy burden on Betty. "She was the primary caretaker," Peggy remembered. "And he would wander. We lived on a busy street. It's scary, to watch your partner have an illness like this."

The Shannons entered an Alzheimer's study at a local hospital, with Betty serving as the control. On whether Shannon knew what was happening to him, Peggy remarked, "There were days he knew, days he didn't. . . . All I can say is when I saw him, there were times it was the same old dad; there were times when he was really out of it." Watching him go, she said simply, was "mostly heartbreaking."

In too-brief moments, the family was given a flash of the Claude they knew. Peggy remembered that she "actually had a conversation with him in 1992 about . . . graduate school programs and what problems I might pursue. And I remember being just amazed how he could cut to the core of the questions I was thinking about, I was like, 'Wow, even in his compromised state he still has that ability.'"

But they were only momentary beams of light in the thickening fog. Within a few years, Shannon's gaps became more pronounced, his moments of clarity fewer and further between. In 1993, remembered Robert Fano, "I asked him something about the past, nothing technical or mathematical, and Claude just answered, 'I don't remember.'" It is one of the cruelties of Shannon's life that a disease of the mind was the cause of his decline. Friends and loved ones lamented that fact nearly as much as they lamented the reality that he would be gone soon.

There was also the more acute unfairness in the fact that, just after he was diagnosed, the digital world he had helped to bring about came into full flower. "Oddly enough, I don't think he even realized what it turned into. . . . He would have been absolutely astounded," Betty said. And he would surely have been pleased by the 1993 announcement of codes whose speed finally approached, but did not break, the Shannon Limit, had the news found any purchase on him.

From 1983 to 1993, Shannon continued to live at Entropy House and carry on as well as he could. Perhaps it says something about the depth of his character that, even in the last stages of his decline, much of his natural personality remained intact. "The sides of his personality that seemed to get stronger were the sweet, boyish, playful sides. . . . We were lucky," Peggy noted. The games and tinkering continued, if at a more measured pace. Arthur Lewbel recalled one of his last interactions with Shannon:

> The last time I saw Claude, Alzheimer's disease had gotten the upper hand. As sad as it is to see anyone's light slowly fade, it is an especially cruel fate to be suffered by a genius. He vaguely remembered I juggled, and cheerfully showed me the juggling displays in his toy room, as if for the first time. And despite the loss of memory and reason, he was every bit as warm, friendly, and cheerful as the first time I met him.

In 1993, Shannon fell, broke his hip, and had to be hospitalized. The cycle of rehabilitation and acute care was lengthy, a grueling period in which the question of what would come next was foremost in the minds of the family. For Betty, Claude was to stay at home—"home was a real refuge for her," Peggy said—and she began to prepare one of the rooms of Entropy House, outfitting it with a hospital bed and other essentials. But Betty herself was getting older, and her daughter had a sense that the challenge of caring for her father might be too much to handle. She urged her mother to consider an assisted living facility, and though she left the final decision up to her, she was relieved when her mom agreed to move Claude to Courtyard Nursing Care Center, roughly three miles from Winchester.

For Betty, her husband's distance from the house changed nothing: she remained utterly committed to him and his care, and she visited twice each and every day. Her daughter, Peggy, was moved to observe that "she was very devoted. She wanted to make sure he had the proper care. She also missed him. He had been the center of her life and that continued even when he moved into the home." For Claude, these visits became treasured events. Betty said, "at noon I'd go over and nurses would line up on a bench waiting for me because when I came in his face would light up and he would grin. I thought that was pretty good."

Other family visited from time to time, and the nursing home staff offered him simple arithmetic problems to help him pass the time. Shannon also remained a tinkerer through even his last days; he set to taking apart his walker, imagining a better design all the way. "He still liked to take things apart and figure how they worked," Betty said. He had maintained use of his body and hands, and he tapped his fingers to music. "He was ambulatory . . . and he could walk around and see the rest of the place and see what was going on, but certainly, his mind was not there." But his being able to move and function at a minimal level proved a bit risky. "They had to be careful with him because he would try the staircase in the nursing home, and go walking down the staircase, even though he had a walker with him. He would go outside and they had to go looking for him."

Ultimately, his movements became impaired, and once-simple things—talking, feeding himself—became difficult. Claude Shannon died on February 24, 2001. His brain was donated for Alzheimer's research. His funeral was held at Lane Funeral Home in Winchester, a small affair.

Years earlier, Shannon had set his mind to the question of this funeral—and imagined something very different. For him, it was an occasion that called for humor, not grief. In a rough sketch, he outlined a grand procession, a Macy's–style parade to amuse and delight, and to sum up the life of Claude Shannon. The clarinetist Pete Fountain would lead the way, jazz combo in tow. Next in line: six unicycling pallbearers, somehow balancing Shannon's coffin (labeled in the sketch as "6 unicyclists/1 loved one"). Behind them would come the "Grieving Widow," then a juggling octet and a "juggling octopedal machine." Next would be three black chess pieces bearing $100 bills and "3 rich men from the West"—California tech investors—following the money. They would march in front of a "Chess Float," atop which British chess master David Levy would square off in a live match against a computer. Scientists and mathematicians, "4 cats trained by Skinner methods," the "mouse group," a phalanx of joggers, and a 417-instrument band would bring up the rear.

This proved impractical. And his family, understandably, preferred a tamer memorial. Shannon was laid to rest in Cambridge, along the Begonia Path in Mount Auburn Cemetery.

Yet in a cemetery home to Supreme Court justices, governors, presidents of universities, and many other famed thinkers, statesmen, and scientists, Shannon's gravestone is unique. An unsuspecting visitor would see SHANNON engraved in pale gray marble and move on. What's concealed, however, is a message on the reverse: covered by a bush, the open section of the marble on the back of the tombstone holds Shannon's entropy formula. Shannon's children had hoped the formula would grace the front of the stone; their mother thought it more modest to engrave it on the back.

And so Claude Shannon's resting place is marked by a kind of code: a message hidden from view, invisible except to those looking for it.

1

Vol. 5. No 4 Gaylord, Mich

SHANNON---WOLF NUPTIALS

Wedding Took Place at Lansing on Wednesday—Date Had Been Kept A Profound Secret

MRS. CLAUDE E. SHANNON

CLAUDE E. SHANNON

Born in 1862 in New Jersey, Claude Shannon's father, Claude Elwood Shannon Sr., worked as a furniture salesman, funeral director, and probate judge. Claude's mother, born Mabel Wolf, was the daughter of a German immigrant; she was a teacher and school principal. Their 1909 wedding announcement was front-page news in Gaylord, testament both to the town's smallness and the Shannons' active roles in the community.

2

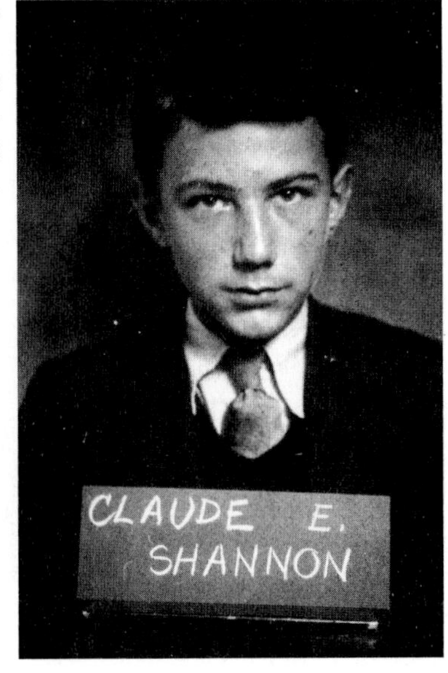

By the time Claude Shannon took this registration photo for the University of Michigan, he was already an accomplished tinkerer. His creations included a makeshift elevator, a backyard trolley, and a telegraph system that sent coded messages along a barbed-wire fence.

3

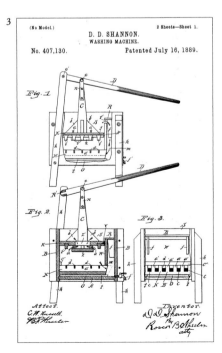

(No Model.) 2 Sheets—Sheet 1.

D. D. SHANNON.

WASHING MACHINE.

No. 407,130. Patented July 16, 1889.

Fig. 1

Fig. 2

Fig. 3

Attest.

Inventor.

Shannon appears to have taken after his grandfather David Shannon Jr., the proud owner of U.S. Patent No. 407,130, a series of improvements on the washing machine. For a boy of Claude Jr.'s mechanical bent, a certified inventor in the family tree was something to brag about.

4 The University of Michigan's College of Engineering had grown dramatically in the years before Shannon's arrival. At one of its public exhibitions, students "surprised their visitors by sawing wood with a piece of paper running at 20,000 revolutions per minute, freezing flowers in liquid air, and showing a bottle supported only by two narrow wires from which a full stream of water flowed—a mystery solved by few." Michigan's engineering buildings were fitted out for heavy industry. They included this foundry . . .

5 

. . . and this naval tank, where students tested the hydrodynamics of model ships.

6 

In the spring of 1934, at the age of seventeen, Claude Shannon claimed his first publication credit, on page 191 of the *American Mathematical Monthly*. Shannon's contribution was the solution to a math puzzle. That he was reading such a journal at all hints at more than the usual attention paid to academic matters; that his solution was chosen points to more than the usual talent.

7

The campus of MIT, where Shannon first made his name as an engineer, was a compromise between architects: a classical dome sitting atop tunnels built for "efficiency and *avoidance of lost motion* by student and teacher, equal to that which obtains in our best industrial works." It was part temple and part factory.

8

At MIT, Shannon joined a team dedicated to managing the differential analyzer: an all-purpose mechanical computer capable of bringing the force of calculus to bear on the industrial problems of power transmission or telephone networks, or on the advanced physics problems of cosmic rays and subatomic particles. The project followed in the footsteps of the wizard-bearded physicist William Thomson, Lord Kelvin, who built an early mechanical computer in 1876.

9

At MIT, Shannon trained as a pilot in his spare time. The professor who taught the flying class urged MIT's president to ban him from taking further lessons: such a brain was too valuable to risk in a crash. The president refused: "Somehow I doubt the advisability of urging a young man to refrain from flying or arbitrarily to take the opportunity away from him, on the ground of his being intellectually superior."

10

The differential analyzer: a brain the size of a room that could whir away at a problem for days and nights on end. "It was a fearsome thing of shafts, gears, strings, and wheels rolling on disks—but it worked."

11

Vannevar Bush was, by most accounts, the most powerful scientist in mid-twentieth-century America. He presided over the differential analyzer at MIT, advised presidents, and led America's scientists through World War II. *Collier's* magazine called him "the man who may win or lose the war"; *Time*, "the general of physics." And not least among these accomplishments was this: he became Claude Shannon's first and most influential mentor.

12

EUGENICS RECORD OFFICE

ESTABLISHED OCTOBER 1, 1910

MRS. E. H. HARRIMAN, FOUNDER

BOARD OF SCIENTIFIC DIRECTORS

ALEXANDER GRAHAM BELL, CHAIRMAN WILLIAM H. WELCH, VICE CHAIRMAN

LEWELLYS F. BARKER, IRVING FISHER, T. H. MORGAN, E. E. SOUTHARD

CHARLES B. DAVENPORT, SECRETARY AND RESIDENT DIRECTOR

H. H. LAUGHLIN, SUPERINTENDENT

THE FUNCTIONS OF THE OFFICE ARE:

1. To serve eugenical interests in the capacity of repository and clearing house.
2. To build up an analytical index of the inborn traits of American families.
3. To train field workers to gather data of eugenical import.
4. To maintain a field force actually engaged in gathering such data.
5. To co-operate with other institutions and with persons concerned with eugenical study.
6. To investigate the manner of the inheritance of specific human traits.
7. To investigate other eugenical factors, such as (a) mate selection, (b) differential fecundity, (c) differential survival, and (d) differential migration.
8. To advise concerning the eugenical fitness of proposed marriages.
9. To publish results of researches.

Traveling to Cold Spring Harbor in the summer of 1939, Shannon arrived at one of the greatest genetics laboratories in America, and one of the greatest scientific embarrassments: the Eugenics Record Office. It housed a treasury of genetic data, which Shannon drew on for his dissertation on theoretical genetics.

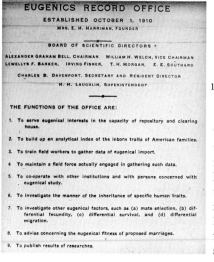

13

Shannon's fellowship at the Institute for Advanced Study in Princeton was prestigious, but less than gratifying: it saw the breakdown of his first marriage and his growing fears about the WWII draft, though it also included a number of run-ins with Albert Einstein.

14

This was the Bell Labs complex as seen from Washington Street in Manhattan's West Village in 1936. "People did very well at Bell Labs," recalled one of Shannon's colleagues, "when they did what others thought was impossible." Shannon signed up for a full-time job when its offices were still in lower Manhattan, near the site of the present-day High Line.

15

Claude Shannon and his colleague Dave Hagelbarger at work at Bell Labs. As another colleague would remark about that time at Bell: "Here I am, I've got the world's knowledge in electrical engineering at my beck and call. All I've got to do is pick up the phone or go see somebody and I can get the answer."

Thornton Fry founded the Labs' math group and assigned Shannon to it. "Mathematicians are queer people," he once said. "That's a fact. So that anybody who was queer enough that you didn't know what to do with him, you said, 'This fellow is a mathematician. Let's have him transferred over to Fry.' "

16

17

Along with Barney Oliver, Shannon and John Pierce (pictured here) formed a genius clique of three at the Labs. As a contemporary joked, "the three of those people were intellectually insufferable."

18

In Manhattan, Shannon was a bachelor—after the end of his brief first marriage—with a small Greenwich Village apartment and a demanding job. He kept odd hours, played music too loud, and relished the New York jazz scene.

 19

As part of the "scientists' war" against Nazi Germany, Shannon contributed to research on cryptography, antiaircraft fire control, and SIGSALY (pictured here), the most ambitious speech-scrambling system of its day.

20 

No one was more influential to Shannon's work on information theory than Ralph Hartley. His 1927 paper on the "Transmission of Information" was the closest approach to date to capturing the essence of information, explaining how scientists might begin to think of it physically rather than psychologically.

 21

Before Shannon's 1948 "A Mathematical Theory of Communication," a century of common sense and engineering trial and error said that noise—the physical world's tax on our messages—had to be lived with. Shannon proved that noise could be defeated, that information sent from Point A could be received with perfection at Point B, not just often, but essentially always. He gave engineers the conceptual tools to digitize information and send it flawlessly, a result considered hopelessly utopian up until the moment Shannon proved it was not.

After 1948, Shannon was heralded by the media as a noteworthy figure in the world of science. He gave television interviews, was written up in national publications, and received honorary degrees.

22

23

In 1948, Claude Shannon struck up a conversation with Betty Moore, a fellow Bell Labs employee, and summoned the courage to ask her out to dinner. That dinner led to a second, the second to a third, until they were dining together every night.

24

Claude and Betty Shannon's courtship was efficient: they met in the fall of 1948, and by early 1949, Claude had proposed—in his "not very formal" way, as Betty recalled. She shared not only his sense of humor but his love of math—the foundation of a partnership that would last the rest of Claude's life.

Betty purchased Claude's first unicycle, sparking his lifelong fascination with the machines. He would build and customize unicycles of his own, ride them through the narrow Bell Labs hallways, and impress visitors with his agility.

25

Even at the height of his scientific celebrity, Shannon remained a tinkerer. His most famous creation, Theseus, was an artificial, maze-solving mouse that could "learn" the location of a metallic piece of cheese. (When the cheese was removed, Theseus could only wander the maze aimlessly. As another scientist observed, "It is all too human.")

26

27

Bell Labs' Murray Hill campus, "where the future, which is what we now happen to call the present, was conceived and designed."

29

Shannon built one of the world's earliest chess-playing machines. Completed in 1949, Shannon's machine handled six pieces and focused on the game's final moves. More than 150 relay switches were used to calculate a move, processing power that allowed the machine to decide within a respectable ten to fifteen seconds.

"I'm a machine and you're a machine, and we both think, don't we?"

30

31

Shannon set four goals for artificial intelligence to achieve by 2001: a chess-playing program that was crowned world champion, a poetry program that had a piece accepted by the *New Yorker*, a mathematical program that proved the elusive Riemann hypothesis, and, "most important," a stock-picking program that outperformed the prime rate by 50 percent. "These goals," he said only half-jokingly, "could mark the beginning of a phase-out of the stupid, entropy-increasing, and militant human race in favor of a more logical, energy conserving, and friendly species—the computer."

32

Norbert Wiener (pictured in the center with Shannon and MIT president Julius Stratton on the left) was a former child prodigy, the inventor of "cybernetics," and the only scientist who could plausibly challenge Shannon's claim to have founded information theory.

33

Claude and Betty purchased a house in Winchester, Massachusetts, a bedroom community eight miles north of MIT. Built in 1858, the house was constructed for Ellen Dwight, a great-granddaughter of Thomas Jefferson. Originally seated on twelve acres, its design was inspired by Monticello.

34

On a trip to Russia, Shannon offered a friendly chess game to a Soviet champion, Mikhail Botvinnik. Botvinnik only took a serious interest in the game when Shannon managed to win the favorable exchange of his knight and a pawn for Botvinnik's rook. After forty-two moves, Shannon tipped his king over, conceding the match. But lasting dozens of moves against Botvinnik, considered among the most gifted chess players of all time, earned Shannon lifelong bragging rights.

35

In 1957, Shannon returned to MIT as a professor. He attracted somewhat less than his share of graduate students: as one observed, "You had to have a pretty big ego to ask someone like Shannon to supervise you!"

 36

On February 6, 1967, President Lyndon B. Johnson presented Claude Shannon with the National Medal of Science in honor of his "brilliant contributions to the mathematical theories of communications and information processing."

 37

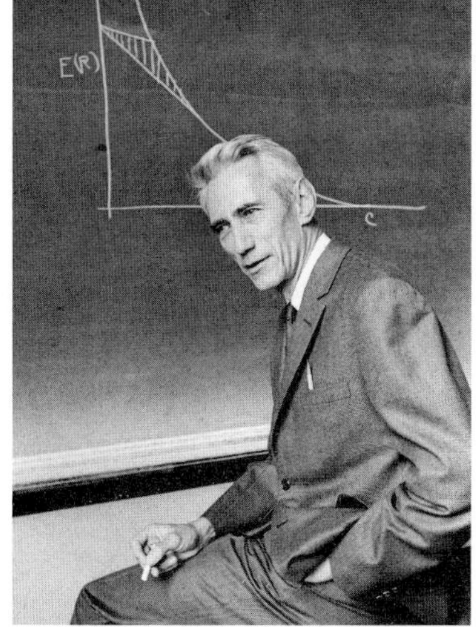

Shannon's early MIT lectures on information theory attracted packed houses, but none drew a bigger crowd than his talk on the stock market, which he delivered to an overflowing crowd in the university's largest lecture hall.

 38

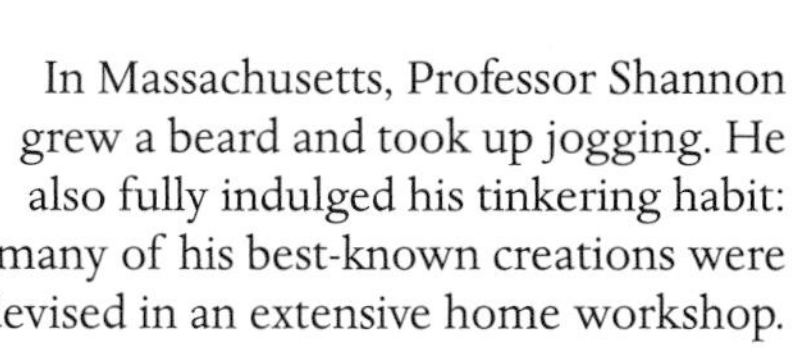

In Massachusetts, Professor Shannon grew a beard and took up jogging. He also fully indulged his tinkering habit: many of his best-known creations were devised in an extensive home workshop.

39

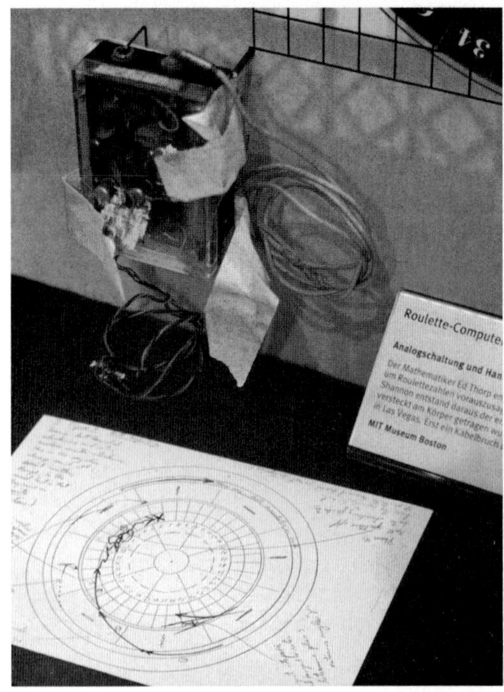

What was arguably world's first wearable computer was developed by Shannon and Edward Thorp to calculate roulette odds. After several successful test runs at Las Vegas casinos, they abandoned the project out of fear of provoking a run-in with the Mafia.

40

Assembled from Shannon's erector set and modeled on W. C. Fields, this robot could bounce-juggle three balls. The balls rebounded off a tom-tom drum, and the robot moved its paddle arms in a rocking motion, "each side making a catch when it rocks down and a toss when it rocks up."

41

Shannon was both a skilled juggler himself and the author of the first serious paper on the mathematics of juggling. He wrote that his readers should "try not to forget the poetry, the comedy and the music of juggling. . . . Does this sound pretentious?"

Shannon kept up his unicycling hobby well into his later years.

42

43

Shannon's was a life spent in pursuit of curious, serious play; he was that rare scientific genius who was just as content rigging up a juggling robot or a flamethrowing trumpet as he was pioneering digital circuits.

44

45

Shannon delighted in the curiosities that grabbed his attention. He could find himself lost in the intricacies of an engineering problem, and then, just as suddenly, become captivated by a chess position. He turned arid and technical sciences into vast and captivating puzzles, the solving of which was play of the adult kind.

32

Aftershocks

The true spirit of delight, the exaltation, the sense of being more than man, which is the touchstone of the highest excellence, is to be found in mathematics as surely as in poetry.

—**Bertrand Russell**

The *New York Times* ran his obituary. Busts and statues were commissioned. A building at the Bell Labs campus was renamed for him. And, then, as far as the general public was concerned, Claude Shannon faded from memory.

His deepest legacy, in some sense, wasn't the one he owned, but the one woven into the work of others—his students, his admirers, later information theorists, engineers, and mathematicians. They kept his memory very much alive, and they did so in the specialist journals in which Shannon made his name. The pens of his fellow engineers and information theorists produced many heartfelt remembrances, recollections that continue even to this day. "A uniquely playful and gentle American genius," one writer noted. "Shannon radiated . . . a powerful inherent intellectual light," wrote another. Yet another writer, who never met Shannon, confessed that at the age of nine he happened upon Shannon's master's thesis—which convinced him to choose a life in mathematics right then and there.

Partly, the reflections were possible because many of the writers had a rare experience in the world of science—that of sharing time on earth with the person who gave birth to their field. But it was more than that: Shannon's work left its lasting mark on generations of American engineers and mathematicians, in part, because it resonated with their fundamental values.

What were those values? Simplicity matters. Elegant math was forceful math. Inessential items, superfluous writing, extra work—all of them should be discarded. In his way of approaching mathematics as an exercise in getting down to the essentials, Shannon produced work that would be regarded as remarkably self-contained, polished, intuitive, and, of course, brilliant—on par with F=ma or $E=mc^2$. A group of Russian mathematicians wrote that in Shannon's work, "the logical and natural development of sections into each other makes an impression that the problem is developing itself." They called this impression the mathematical virtue of "integrity," the sense that a Shannon paper is a seamless whole. Still another of Shannon's contemporaries put it more poetically: "[His] ideas form a beautiful symphony, with repetition of themes and growing power that still form an inspiration to all of us. This is mathematics at its very best."

In 1948, Shannon's theoretical work posed as many questions as it answered. But the value of that challenge shouldn't be underestimated. There was the far-off prospect of the Shannon Limit, which would prove increasingly useful decades later; even now, the limit remains the tantalizing outer edge of communication, a goal that engineers continue to chase. But those were the narrow, practical elements. The striking feature of his paper is the reverberation, the way in which it inaugurated an entire field of study, a body of dialogue and deliberation that would long outlive its author. "It was like an earthquake and the aftershocks haven't finished yet!" observed Anthony Ephremides, an information theorist of a later era. Few papers can claim an impact so enduring (it has more than 91,000 citations and counting!), and it's no exaggeration to say that, though information theory had important antecedents prior to Shannon, the formal study of information begins in earnest with his work. As one writer would put it, many decades later, "For many scientists, Shannon's discovery was the equivalent of waking up and finding marble on their doorsteps."

The marble he unearthed would be carved by others; Shannon's work condemned him, to some extent, to a legacy as an antecedent. He is one of the authors of the information architecture that now binds the planet—but he will likely never approach the name recognition of a Steve Jobs or Bill Gates. Beyond his own aversion to such attention, his anonymity could be chalked up to the distance between his work and the technologies we use every day. When a world-class engineer says that "all the advanced signal processing that enables us to send high-speed data was done as an outgrowth of Claude Shannon's work on information theory," the statement rings true to people in the know—and means very little to the untrained.

Yet there is value in rethinking Claude Shannon—but not in the way we'd imagine. Consider him not only as a distant forefather of the digital era, but as one of the great creative generalists of the twentieth century: not solely as someone who laid the foundation of the Information Age, but as someone who trained a powerful intellect on topics of deep interest, and continued to do so beyond the point of short-term practicality.

What can we learn from *that* Claude Shannon?

For one thing, Shannon's body of work is a useful corrective to our era of unprecedented specialization. His work is wide-ranging in the best sense, and perhaps more than any twentieth-century intellect of comparable stature, he resists easy categorization. Was he a mathematician? Yes. Was he an engineer? Yes. Was he a juggler, unicyclist, machinist, futurist, and gambler? Yes, and then some. Shannon never acknowledged the contradictions in his fields of interest; he simply went wherever his omnivorous curiosity led him. So it was entirely consistent for him to jump from information theory to artificial intelligence to chess to juggling to gambling—it simply didn't occur to him that investing his talents in a single field made any sense at all.

There were links from field to field, of course. And it goes without saying that Shannon understood the bridges between his work in information theory and his work on robotics and investing and computer chess. Few have had a better intuitive sense of how the information revolution would fundamentally alter our world in all its aspects. But that sense led Shannon to choose exploration rather than specialization. He could have continued to trade on the success of information theory for decades. By the time he arrived at MIT, though, his attention was elsewhere. Students of that era recall that Shannon himself just didn't seem especially engaged by information theoretic questions and problems; however, if you brought him something in robotics or artificial intelligence, they recalled, his ears perked up and he paid special attention.

The great Russian mathematician Andrey Kolmogorov put it like this in 1963:

> In our age, when human knowledge is becoming more and more specialized, Claude Shannon is an exceptional example of a scientist who combines deep abstract mathematical thought with a broad and at the same time very concrete understanding of vital problems of technology. He can be considered equally well as one of the greatest mathematicians and as one of the greatest engineers of the last few decades.

This indifference to seeming contradictions extended to the way he lived his life. He had the option of worldwide fame, yet he preferred to remain largely anonymous. He wrote pathbreaking papers, then, unsatisfied with their present state, postponed them indefinitely in favor of more pressing curiosities. He made himself wealthy by studying the movements of markets and the potential of start-ups, yet he lived with conspicuous modesty. He reached the heights of the ivory tower, with all the laurels and professorial chairs to prove it, but felt no shame playing games built for children and writing tracts on juggling. He was passionately curious, but also, at times, unapologetically lazy. He was among the most productive, honored minds of his era, and yet he gave the appearance that he would chuck it all overboard for the chance to tinker in his gadget room.

His style of work was characterized by such lightness and levity, in fact, that we can sometimes forget the depth and difficulty of the problems he took on. For all the fun he was having, Shannon tackled some of the most significant scientific questions of his era and worked at the boundaries of math, computer science, and engineering—in some cases helping to firm up what the boundaries were! As AI pioneer Marvin Minsky said on the news of Shannon's death, "For him, the harder a problem might seem, the better the chance to find something new."

Such an approach took courage—a quality about Shannon captured in the words of one of his Bell Labs office mates, Richard Hamming. In a now-famous talk called "You and Your Research," Hamming outlines for a group of students the attributes that make for first-rate research in mathematics and other disciplines. He singles out Shannon for special mention, and notes that part of what gave Shannon's work force was his bravery:

> Courage is one of the things that Shannon had supremely. You have only to think of his major theorem. He wants to create a method of coding, but he doesn't know what to do so he makes a random code. Then he is stuck. And then he asks the impossible question, "What would the average random code do?" He then proves that the average code is arbitrarily good, and that therefore there must be at least one good code. Who but a man of infinite courage could have dared to think those thoughts? That is the characteristic of great scientists; they have courage. They go forward under incredible circumstances; they think and continue to think.

We don't usually associate the fields of mathematics or engineering with the ancient virtue of courage. But Shannon's wasn't the usual contribution to those fields, either, and though he surely would have been the last to admit this, it took a great deal of daring to think as Shannon thought and to live as Shannon lived. All of this also had an effect on those around him, including his students. "When you work with someone like Shannon, you expand your horizons, you try to reach far," remarked Len Kleinrock.

Importantly, his courage was joined to an ego so self-contained and self-sufficient that it looked, from certain angles, like the absence of ego. This was the keystone quality of Shannon, the one that enabled all the others. At almost every opportunity for self-promotion, Shannon demurred. Mathematicians worry about spending time on problems of insufficient difficulty, what they derisively call "toy problems"; Claude Shannon worked with *actual* toys in public! Time and again, he pursued projects that might have caused others embarrassment, engaged questions that seemed trivial or minor, then managed to wring the breakthroughs out of them. It takes no small degree of self-assurance to try to build a brain that would outclass one's own—or, for that matter, to build a machine whose only function is to turn itself off.

And that is connected, we think, to the other great hallmark of Shannon's life: the value of finding joy in work. We expect our greatest minds to bear the deepest scars; we prefer our geniuses tortured. But with the exception of a few years in his twenties when Shannon passed through what seems like a moody, possibly even depressive, stage, his life and work seemed to be one continuous game. He was, at once, abnormally brilliant and normally human.

He did none of this consciously; he wasn't straining to give the appearance of fun. Shannon simply delighted in the various curiosities that grabbed his attention, and the testimony of those around him suggests that it was a delight that, like his mind, was polymorphous. He could find himself lost in the intricacies of an engineering problem, and then, just as suddenly, become captivated by a chess position. He had a flair for the dramatic and the artistic; we see it in the flaming trumpet, Theseus the mouse, a flagpole he hand-carved out of an oversize tree on his property, the juggling clowns he built to exacting specifications. Shannon's admirers are just as quick to compare him to M. C. Escher or Lewis Carroll as they are to put him in the company of

Albert Einstein or Isaac Newton. He turned arid and technical sciences into vast and captivating puzzles, the solving of which was play of the adult kind. It says something about Claude Shannon and his instinct for play that his work found its way into both the pages of journals and the halls of museums.

In one sense, it may be impossible to draw anything from this. Shannon's enjoyment seems sui generis. But perhaps his example can still remind us of the vast room for lightness in fields usually discussed in sober tones. These days it's rare to talk about math and science as opportunities to revel in discovery. We speak, instead, about their practical benefits—to society, the economy, our prospects for employment. STEM courses are the means to job security, not joy. Studying them becomes the academic equivalent of eating your vegetables—something valuable, and state sanctioned, but vaguely distasteful.

This seems, at least to us, not as Shannon would have wanted it. Shannon was an engineer—a man more attuned to practicality than most—and yet he was drawn to the idea that knowledge was valuable for its own sake and that discovery was pleasurable in its own right. As he himself put it, "I've been more interested in whether a problem is exciting than what it will do." One of his contemporaries, remarking on the peculiarity of a world-class mathematician with a serious interest in unicycles, put Shannon's love of these strange machines specifically, as well as his passions generally, in perspective: "He was not interested in forming a company to build unicycles. He was interested in finding out what made unicycles fun and finding out more about them."

And his approach inspired a generation of remarkable innovation. Consider the words of Bob Gallager, describing the mood of the minds working on information theory around the same time as Shannon:

> Shannon's puzzle-solving research style was in full swing when I was an MIT graduate student. Intellectualism was in the air. Everyone wanted to understand mathematics and physics as well as communication. Starting companies, making millions, developing real applications was secondary. There was interest in bringing the theory closer to reality, but it was theory-based. Our role models were relaxed, curious, and had time to reflect.

We might be hard-pressed to find an academic department today fitting that description—but surely it is a worthwhile ambition.

Toward the end of his life, Shannon still maintained his cheekiness, his insouciance toward even the highest of the highbrow. After promising *Scientific American* his article on the physics of juggling, his attention drifted—and he chanced on an entirely unrelated project. From that came this note to his editor, written in 1981:

> *Dear Dennis:*
>
> *You probably think I have been fritterin', I say fritterin', away my time while my juggling paper is languishing on the shelf. This is only half true. I have come to two conclusions recently:*
>
> *1) I am a better poet than scientist.*
> *2) Scientific American should have a poetry column.*
>
> *You may disagree with both of these, but I enclose "A Rubric on Rubik Cubics" for you.*
>
> *Sincerely,*
> *Claude E. Shannon*
>
> *P.S. I am still working on the juggling paper.*

What followed was a seventy-line poem on the subject of Rubik's cubes, "sung to 'Ta-Ra-Ra! Boom-De-Ay!' (with an eight bar chorus)" and complete with footnotes. And it was clear from the rhyme and rhythm that the author had spent time playing with the words on his tongue, rearranging them in his head, singing them aloud to himself. The project was seriously unserious.

And the juggling article? Like so many artifacts of Shannon's mind, it acquired dust. Shannon's attention had shifted, and whatever he needed to say about juggling had been said, at least to his satisfaction. He did, however, have one regret about the episode. He was disappointed that his poem never made it into the pages of *Scientific American.*

Laughing, he declared, "That's one of my great works!"

Acknowledgments

This book could have been written in two ways: downward, or upward. A book like this is written downward when it's the work of an expert, straining to send a decipherable message to the rest of us without dumbing down, struggling to remember what it must have been like to be a novice. A book like this is written upward when it's the work of learners, struggling to communicate what they are learning, as part of the very process of learning it. Books of the first kind come from the satisfaction of already knowing. Books of the second kind come from what physicist and bon vivant Richard Feynman called the pleasure of finding things out.

Each model has its attractions and its failings, but this has been a book of the second kind, a book written upward. We are biographers, not mathematicians or physicists or engineers. The best we can say for this inexpert book of ours is that we've tried to write as we'd like to live. That is, we began with a nagging sense that there is something harmful in using without understanding, or at least trying to understand. We began with the idea that there is something ungrateful and grasping in enjoying our bounty of information without bothering to understand how it got here.

We are not the first to live with that sense, nor to try to remedy it. Here is how Arthur Koestler, a physics student turned novelist, once put it:

> Modern man lives isolated in his artificial environment, not because the artificial is evil as such, but because of his lack of

> comprehension of the forces which make it work—of the principles which relate his gadgets to the forces of nature, to the universal order. It is not central heating which makes his existence "unnatural," but his refusal to take an interest in the principles behind it. By being entirely dependent on science, yet closing his mind to it, he leads the life of an urban barbarian.

We would add: it is not the Internet that is unnatural, nor our feast of information, but a refusal to consider what their origins are, how and why they are here, where they sit in the flow of our history, and what kinds of men and women brought them about. We think there is something of an obligation in beginning to learn these things. We think that the honor our subject would have cared about—if he cared at all—would not have been adulation, but a bit of comprehension.

We've had great help in fulfilling this obligation. Dan Kimerling is a friend and entrepreneur—and someone who, no doubt, Shannon would have admired. He first suggested the idea of a biography of Claude Shannon. It may have been the casual remark of a friend (accompanied by a book about Bell Labs), but it has led to the work you hold in your hands. So for that inspiration, we are grateful to Dan.

For believing in us from the start and for seeing the value in this project when it was simply an idea in our minds, we are indebted to Laura Yorke, our agent. She encouraged us to pitch only the faintest of ideas to none other than Alice Mayhew and prodded us at the right times to bring this book to pass. She is a legend in the book business and deservedly so!

No one has a keener sense of a book's possibilities than Simon & Schuster's Alice Mayhew. She, too, is legendary, and now we understand why. Among the many blessings of the last few years, the foremost is having her as the editor of this book. She had boundless faith in this work, even and especially in the moments when our own faith wavered. As she has for a generation of biographers, she brought out our best. For that, and for her peerless editing, we are eternally grateful. We are grateful as well to Stuart Roberts on her team for all he did to make this project a success. Stuart is as smart, patient, and kind as they come, and there's good reason why he is Alice Mayhew's right hand.

Jon Gertner, author of *The Idea Factory,* inspired this project unknowingly and then collaborated on it generously. He responded to queries consequential and not, offered contacts and research, and shared an unpublished oral history of Thornton Fry and the Bell Labs mathematics group that proved critically useful. He was that most valuable guide for a writer of historical narrative: someone who has walked the same path before and knows where the dead ends are. He was generous enough to point them out to us, and we are deeply grateful for his kindness. (Oh, and if the reader hasn't already read his book *The Idea Factory,* we strongly recommend it. There is not a better history of Bell Labs, and few better portraits of how innovative organizations come to be built.) Also invaluable to us as guides to Shannon's life and work were James Gleick's *The Information* and Erico Marui Guizzo's master's thesis, "The Essential Message."

Johannah-King Slutzky proved a first-rate research assistant. Finding people outside of the worlds of science and engineering who know Shannon's name is difficult; finding someone like Johannah—who had written her own piece about Shannon before she started working on this project—was a real stroke of luck. She was diligent, thoughtful, and as excited about the research as we were.

Professor Sergio Verdú, of Princeton University, was an indispensable guide to the world of information theory and brought his collegiality and thoughtfulness to our every interaction. His enthusiasm for the arcana of Shannon's life kept us motivated throughout the project, and he devotedly read every page and fixed many errors. As we write this, he and filmmaker Mark Levinson (of *Particle Fever* fame) are at work on a documentary on Shannon. We have no doubt that it will be marvelous.

We connected with Dr. Alex Magoun late in the life cycle of this project—but boy, are we glad we did! He provided his own copyedit and corrected many errors in the draft. That he read with both an enthusiast's eye for the subject and a trained historian's sense for where we might have erred helped us immeasurably. We are grateful for the many hours he spent and the many mistakes he caught.

Marcus Weldon and the entire Nokia Bell Labs team—including and especially Peter Winzer and Ed Eckert—opened their doors and archives to us. We are grateful for the time they spent and the resources they shared. Understanding Bell Labs is crucial to making sense of Shannon's life, and we could not have done this work without them.

Will Goodman, Internet sleuth extraordinaire, aided us in tracking down the contact information of various Shannon family members and contemporaries. We like to think Shannon himself might have been impressed by Will's roving curiosity—he's a twenty-first-century tinkerer-detective if ever there was one.

To the Shannon family, we thank you for taking the time to share family lore with two perfect strangers. Betty Shannon agreed to speak with us, a conversation that gave us a rich look at the relationship between her and her late husband. Claude's son and daughter, Andrew and Peggy Shannon, also spoke to us at length. Both Shannons were generous enough to read a familiar story about a man they knew well and correct errors (and even typos!) along the way. We could not have completed this project without their help, and we are in their debt.

Like the Shannons, many people answered cold emails and phone calls and took the time to sit down with us and talk. Robert Gallager allowed us an interview—and then painstakingly read each page of the draft and corrected numerous missteps. He was incredibly generous with his time and with his patience for two nontechnical writers. Arthur Lewbel also read a draft, offered wise suggestions, and gave us a window into Shannon's life as a juggler that we could not have gotten otherwise. Tom Kailath helped us make sense of Shannon and Wiener's contributions to information theory and read an early version of the text. Dave Forney wrote a long memo about this text that proved enormously helpful and helped us lock down the mathematics as best as two nonmath minds could. We appreciate their time and contribution.

Kevin Currie helped us find, assemble, and select the photography in the middle of this book. With no prior experience doing such a thing, he jumped right in, and he did a marvelous job. Without his help, we could not have gathered the necessary photos to tell this story well—and we are grateful to him for his help.

Our deepest thanks also to Brockway McMillan, Irwin Jacobs, Ronald and Fan Chung Graham, John Horgan, Larry Roberts, Anthony Ephremides, Maria Moulton-Barrett, Len Kleinrock, Henry Pollak, Norma Barzman, Ed Thorp, Martin Greenberger, the late Bob Fano, and the late Solomon Golomb. At one point or another, they indulged us with their time and help, and this project is richer for their involvement.

And to our own families, this should bring to a close the sharing of endless Shannon trivia. No, seriously, we've reached our limit, and we'll stop. Though for babies Venice and Abigail, born within a week of one another during the course of this project's final year, there may be a bit more information to come.

Notes

Epigraph

ix *"Geniuses are the luckiest"*: W. H. Auden, "Foreword," in Dag Hammarskjöld, *Markings* (New York: Knopf, 1964), xv.

Introduction

xi *"This is—ridiculous"*: Anthony Ephremides, "Claude E. Shannon 1916–2001," *IEEE Information Theory Society Newsletter*, March 2001.

xi *"as if Newton had showed up"*: John Horgan, "Claude E. Shannon: Unicyclist, Juggler, and Father of Information Theory," *Scientific American*, January 1990, 22B.

xi *"the Magna Carta"*: Ibid., 22A.

xii *"a bomb"*: John Pierce, "The Early Days of Information Theory," *IEEE Transactions on Information Theory* 19, no. 1 (1973): 4.

xii *"a fearsome thing"*: Philip McCord Morse, *In at the Beginnings: A Physicist's Life* (Cambridge, MA: MIT Press, 1977), 121.

xiii *"became the basic concept"*: Walter Isaacson, *The Innovators: How a Group of Inventors, Hackers, Geniuses, and Geeks Created the Digital Revolution* (New York: Simon & Schuster, 2014), 49.

xiii *"possibly the most"*: Howard Gardner, quoted in "MIT Professor Claude Shannon Dies; Was Founder of Digital Communications," *MIT News*, February 27, 2001, newsoffice.mit.edu/2001/shannon.

xiii *"the operations of genius"*: Harold Arnold, quoted in Jon Gertner, *The Idea Factory: Bell Labs and the Great Age of American Innovation* (New York: Penguin, 2012), 121.

xiii *"People did very well"*: Henry Pollak, interviewed by the authors August 7, 2014.

xiii *"an analysis of some of the fundamental properties"*: Letter from Claude Shannon to Vannevar Bush, February 16, 1939, Claude Elwood Shannon Papers, Library of Congress.

xiv *"How he got that insight"*: Robert Fano, quoted in W. Mitchell Waldrop, "Claude Shannon: Reluctant Father of the Digital Age," *MIT Technology Review*, July 1, 2001, www.technologyreview.com/s/401112/claude-shannon-reluctant-father-of-the-digital-age.

xiv *"a mote of dust"*: Carl Sagan, *Pale Blue Dot: A Vision of the Human Future in Space* (New York: Random House, 1994), 6.

xiv *"bandwagon"*: Claude Shannon, "The Bandwagon," *IRE Transactions—Information Theory* 2, no. 1 (1956): 3.

xv "XFOML RXKHRJFFJUJ": Claude Shannon, "A Mathematical Theory of Communication," in *Claude Elwood Shannon: Collected Papers*, ed. N. J. A. Sloane and Aaron D. Wyner (New York: IEEE Press, 1992), 14. The paper was originally printed in *Bell System Technical Journal* 27 (July, October 1948): 379–423, 623–56.

Chapter 1: *Gaylord*

3 *"53‡‡†"*: Edgar Allan Poe, "The Gold-Bug," in *The Gold-Bug and Other Tales*, ed. Stanley Appelbaum (Mineola, NY: Dover, 1991), 100.

4 *leather straps*: Delbert Trew, "Barbed Wire Telegraph Lines Brought Gossip and News to Farm and Ranch," *Farm Collector*, September 2003. See also David B. Sicilia, "How the West Was Wired," *Inc.*, June 1997.

5 *"Shannon-Wolf Nuptials"* . . . "If anything is wanted": *Otsego County Times*, August 27, 1909.

5 *"wedding gown of white satin"*: *Otsego County Times*, August 27, 1936. The reference is to the wedding of Mabel's daughter, at which she wore her mother's dress.

5 *"Mr. Shannon, the groom"*: *Otsego County Times*, August 27, 1909.

6 *"Something which should be found"*: C. E. Shannon, advertisement, *Otsego County Herald and Times*, November 1, 1912.

6 *"He would sometimes help me"*: Shannon, interviewed by Donald J. Albers, 1990.

7 *"glowing recommendations"*: Ibid.

7 *"At a meeting"*: Reprinted in "A Brief History of Gaylord Community Schools—1920 to 1944," *Otsego County Herald Times*, May 2, 1957, goo.gl/oVb0pT.

7 *Library Board*: "Mrs. Mabel Shannon Dies in Chicago," *Otsego County Herald Times*, December 27, 1945.

8 *"The fact that"*: Perry Francis Powers and H. G. Cutler, *A History of Northern Michigan and Its People* (Chicago: Lewis, 1912), iv.

8 *ten pins*: H. C. McKinley, "Step Back in Time: A New County Seat and the First Newspaper," *Gaylord Herald Times*, reprinted January 6, 2016, www.petoskeynews.com/gaylord/featured-ght/top-gallery/step-back-in-time-a-new-county-seat-and-the/article_88155b9f-0965-56c2-b2ce-85456427fc70.html.

8 *"WISCONSIN GIRL," "A woman smoking" . . . "LUMBERJACK DIES" . . . "VERN MATTS LOSES FINGER," "MEETING CALLED TO DISCUSS ARTICHOKES" . . . "splotches of silver"*: *Otsego County Herald and Times*, April 28, 1916; September 20, 1923; September 27, 1923; *Otsego County Herald Times*, November 18, 1926; September 27, 1928; September 13, 1928; *Otsego County Herald and Times*, September 27, 1923.

8 *"the first business"*: *Otsego County Herald Times*, October 22, 1969.

9 *"Norbert always felt"*: Paul A. Samuelson, "Some Memories of Norbert Wiener," in *The Legacy of Norbert Wiener: A Centennial Symposium in Honor of the 100th Anniversary of Norbert Wiener's Birth, October 8–14, 1994, Massachusetts Institute of Technology, Cambridge, Massachusetts*, ed. David Jerison, I. M. Singer, and Daniel W. Stroock (Providence, RI: American Mathematical Society, 1997), 38.

9 *"one of Gaylord's most popular girls"*: *Otsego County Times*, August 27, 1936.

9 *"She was a model student"*: Shannon, interviewed by Albers, 1990.

9 *"A Poor Boy"*: Reprinted in "A Brief History of Gaylord Community Schools—1920 to 1944," *Otsego County Herald Times*, May 2, 1957, goo.gl/oVb0pT.

10 *"Some names" . . . "boys in those grades"*: Letter from Shannon to Virginia Howe, May 2, 1983, Shannon Papers.

10 *"I think one"*: Claude Shannon, interviewed by Friedrich-Wilhelm Hagemeyer, February 28, 1977.

10 *"always seems a little dull"*: Letter from Shannon to Irene Angus, August 8, 1952, Shannon Papers.

10 *"first place"*: *Otsego County Herald Times*, April 17, 1930.

11 *"dirt, settlings, and foul matter"*: U.S. Patent No. 407,130.

11 *"As a young boy"*: Shannon, interviewed by Hagemeyer, February 28, 1977.

11 *"He and my brother"*: Quoted in Julie Kettlewell, "Gaylord Honors 'Father to the Information Theory,'" *Otsego Herald Times*, September 3, 1998.

11 *"Claude was the brains"*: Quoted in Melinda Cerny, "Engineering Industry Honors Shannon, His Hometown," *Otsego Herald Times*, September 3, 1998.

11 *John Ogden*: Jack Harpster, *John Ogden, The Pilgrim (1609–1682): A Man of More than Ordinary Mark* (Cranbury, NJ: Associated University Presses, 2006), 209.

Chapter 2: *Ann Arbor*

13 *"8. Have you earned"*: Claude Shannon Alumnus File, Bentley Historical Library, University of Michigan.

14 *"the century to come"*: Quoted in James Fraser Cocks and Cathy Abernathy, *Pictorial History of Ann Arbor, 1824–1974* (Ann Arbor: Michigan Historical Collections/Bentley Historical Library Ann Arbor Sesquicentennial Committee, 1974), 54.

14 *"I am not at all"*: Ibid., 92.

15 *"enrollments"*: Anne Duderstadt, "Engineering," in *The University of Michigan: A Photographic Saga*, umhistory.dc.umich.edu/history/publications/photo_saga/media/PDFs/12%20Engineering.pdf.

15 *"with his characteristic chuckle"*: Hillard A. Sutin, "A Tribute to Mortimer E. Cooley," *Michigan Technic*, March 1935, 103.

15 *"Gentlemen, if you could"*: Ibid., 105.

16 *"surprised their visitors"*: *Michigan Alumnus* 22 (1916): 463.

16 *"I wasn't really quite sure"*: Shannon, interviewed by Hagemeyer, February 28, 1977.

17 *"A feature of all meetings"*: F. W. Owens and Helen B. Owens, "Mathematics Clubs—Junior Mathematics Club, University of Michigan," *American Mathematical Monthly* 43, no. 10 (December 1936): 636.

17 *"Claude Shannon has been made"*: "Gaylord Locals," *Otsego County Herald Times*, November 15, 1934.

17 *"Breakfast at the dining hall"*: *1934 Michiganensian*, ed. C. Wallace Graham et al. (Ann Arbor, 1934), 364.

18 *"removes his legs"*: Ibid., 370–72.

18 *"He laughed in small explosions"*: Maria Moulton-Barrett, *Graphotherapy* (New York: Trafford, 2005), 84.

19 *first publication credit*: Shannon, "Problems and Solutions—E58," *American Mathematical Monthly* 41, no. 3 (March 1934): 191–92.

19 *"problems believed to be new"*: Otto Dunkel, H. L. Olson, and W. F. Cheney, Jr., "Problems and Solutions," *American Mathematical Monthly* 41, no. 3 (March 1934): 188–89.

19 *"In the following division"*: R. M. Sutton, "Problems for Solution," *American Mathematical Monthly* 40, no. 8 (October 1933): 491.

19 *"In two concentric circles"*: G. R. Livingston, "Problems for Solution," *American Mathematical Monthly* 41, no. 6 (June 1934): 390.

20 *after his undergraduate days were over*: With his MIT admission in hand, Shannon would spend the summer of 1936 in Ann Arbor, "taking extracurricular work"; *Otsego County Herald Times*, August 6, 1936.

20 *"I pushed hard for that job"*: Shannon, interviewed by Albers, 1990.

Chapter 3: *The Room-Sized Brain*

22 *"it ran over"*: Quoted in G. Pascal Zachary, *Endless Frontier: Vannevar Bush, Engineer of the American Century* (Cambridge, MA: MIT Press, 1999), 26.

22 *"the man who" . . . "the general of physics"*: J. D. Ratcliff, "Brains," *Collier's*, January 17, 1942; "Vannevar Bush: General of Physics," *Time*, April 3, 1944.

22 *"an apple drops from a tree"*: Vannevar Bush, *Pieces of the Action* (New York: Morrow, 1970), 181.

24 *"the towing of one car"*: Harold Hazen, quoted in David A. Mindell, *Between Human and Machine: Feedback, Control, and Computing before Cybernetics* (Baltimore: Johns Hopkins University Press, 2002), 151.

25 *"You and I"*: Silvanus P. Thomson, *The Life of William Thomson, Baron Kelvin of Largs* (London: Macmillan, 1910), 1:98.

25 *"Go, wondrous creature"*: Alexander Pope, "An Essay on Man," 2.19–20; epigraph to William Thomson, "Essay on the Figure of the Earth," Kelvin Collection, University of Glasgow.

26 *"calculation of so methodical a kind"*: Thomson, "The Tides: Evening Lecture to the British Association at the Southampton Meeting, August 25, 1882," in *Scientific Papers*, ed. Charles W. Eliot (New York: Collier & Son, 1910), 30:307.

27 *"I would construct a machine"*: Quoted in A. Ben Clymer, "The Mechanical Analog Computers of Hannibal Ford and William Newell," *IEEE Annals of the History of Computing* 15, no. 2 (1993): 23.

27 *Hannibal Ford was not the first*: For instance, an earlier integrator was developed by Arthur Pollen for the British navy but was not widely adopted. See Norman Friedman, *Naval Firepower: Battleship Guns and Gunnery in the Dreadnought Era* (Barnsley, England: Seaforth, 2008), 53ff.

27 *"a marvel of precision"*: Quoted in Karl L. Wildes and Nilo A. Lindgren, *A Century of Electrical Engineering and Computer Science at MIT, 1882–1982* (Cambridge, MA: MIT Press, 1986), 87.

28 *"By turning the nut"*: Daniel C. Stillson, U.S. Patent 126,161,

"Improvement in Pipe Wrenches," U.S. Patent Office, 1872 (an example of a patent that Bush may have used for this exercise).

29 *"A man learns"*: John Perry, *The Calculus for Engineers* (London: Edward Arnold, 1897), 5, cited in Larry Owens, "Vannevar Bush and the Differential Analyzer: The Text and Context of an Early Computer," *Technology and Culture* 27, no. 1 (1986): 63–95.

29 *"well-stocked with clay"*: Benchara Branford, *A Study of Mathematical Education* (Oxford: Clarendon, 1908), viii, cited in Owens, "Vannevar Bush and the Differential Analyzer."

29 *"uncertain always"*: Paul Fussell, *Class: A Guide Through the American Status System* (New York: Touchstone, 1992), 64.

30 *"It was a fearsome thing"*: Morse, *In at the Beginnings*, 121.

30 *"in calculating the scattering"*: D. R. Hartree, "The Bush Differential Analyzer and its Applications," *Nature* 146 (September 7, 1940): 320.

31 *"still interpreted mathematics"*: Owens, "Vannevar Bush and the Differential Analyzer," 95.

31 *"It is an analogue machine"*: Quoted in Zachary, *Endless Frontier*, 49.

Chapter 4: *MIT*

32 *"Institute folklore"*: Fred Hapgood, *Up the Infinite Corridor: MIT and the Technical Imagination* (New York: Basic Books, 1994), 61.

33 *"efficiency and* avoidance of lost motion": John Ripley Freeman, "Study No. 7 for New Buildings for the Massachusetts Institute of Technology," MIT Libraries, Institute Archives and Special Collections, libraries.mit.edu/archives/exhibits/freeman.

34 *"an electrical switch"*: James Gleick, *The Information: A History, a Theory, a Flood* (New York: Pantheon, 2011), 173.

35 *"As a material machine"*: W. E. Johnson, "The Logical Calculus," *Mind: A Quarterly Review of Psychology and Philosophy* 1 (1892): 3.

36 *blue-eyed and left-handed*: For this example, and for our discussion of Boolean logic in general, we are indebted to Paul J. Nahin, *The Logician and the Engineer: How George Boole and Claude Shannon Created the Information Age* (Princeton, NJ: Princeton University Press, 2013), esp. 45–47.

37 *"(x + y)' = x'y'"*: This particular law was one of those identified by Augustus De Morgan, another important contributor to formal logic.

38 *"It's not so much"*: "Profile of Claude Shannon—Interview by Anthony Liversidge," in *Claude Elwood Shannon: Collected Papers*, xxvi.

39 *"I think I had more fun"*: Ibid.

39 *"any circuit"*: Shannon, "A Symbolic Analysis of Relay and Switching Circuits," *Transactions of the American Institute of Electrical Engineers* 57 (1938): 471.

40 *Consider a problem*: This example is derived from INTOSAI Standing Committee on IT Audit, "1 + 1 = 1: A Tale of Genius," *IntoIT* 18 (2003): 56.

42 *"possibly the most important" . . . "One of the greatest" . . . "The most important" . . . "Monumental"*: Gardner, quoted in "MIT Professor Claude Shannon Dies"; Solomon W. Golomb, "Retrospective: Claude E. Shannon (1916–2001)," *Science*, April 20, 2001, 455; William Poundstone, *Fortune's Formula: The Untold Story of the Scientific Betting System That Beat the Casinos and Wall Street* (New York: Hill & Wang, 2005), 20; Marvin Minsky, quoted in *Claude Elwood Shannon: Collected Papers*, xix.

43 *"became the basic concept"*: Isaacson, *The Innovators*, 49.

43 *"an annus mirabilis"*: Ibid., 38.

43 *"an all-or-none device"*: John von Neumann, "First Draft of a Report on the EDVAC," in *The Origins of Digital Computers: Selected Papers*, ed. Brian Randell (New York: Springer-Verlag, 1973), 362.

43 *"Years ago"*: Hapgood, *Up the Infinite Corridor*, 11.

Chapter 5: *A Decidedly Unconventional Type of Youngster*

45 *"Skin Effect Resistance," etc.*: Victor J. Decorte, "Skin Effect Resistance Ratio of a Circular Loop of Wire" (MS thesis, Massachusetts Institute of Technology, 1929), hdl.handle.net/1721.1/81515; Burdett P. Cottrell, "An Investigation of Two Methods of Measuring the Acceleration of Rotating Machinery" (MS thesis, Massachusetts Institute of Technology, 1929), hdl.handle.net/1721.1/85720; R. A. Swan and W. F. Bartlett, "Three Mechanisms of Breakdown

of Pyrex Glass" (BS thesis, Massachusetts Institute of Technology, 1929), hdl.handle.net/1721.1/49611; James Sophocles Dadakis, "A Plan for Remodeling an Industrial Power Plant" (BS thesis, Massachusetts Institute of Technology, 1930), hdl.handle.net/1721.1/51558; Herbert E. Korb et al., "A Proposal to Electrify a Section of the Boston and Maine Railroad Haverhill Division" (MS thesis, Massachusetts Institute of Technology, 1933), hdl.handle.net/1721.1/10560.

46 *"What's your secret"*: "Profile of Claude Shannon—Interview by Anthony Liversidge," in *Claude Elwood Shannon: Collected Papers*, xxxii.

46 *"I don't happen to be"*: Ibid., xxviii.

47 *"I am convinced"*: Letter from R. H. Smith to Karl Compton, April 11, 1939, Office of the President Records, MIT Archive, cited in Erico Marui Guizzo, "The Essential Message: Claude Shannon and the Making of Information Theory" (MS diss., Massachusetts Institute of Technology, 2003), 13.

47 *"Somehow I doubt"*: Letter from Compton to Smith, April 13, 1939, Office of the President Records, MIT Archive, cited in Guizzo, "The Essential Message," 13.

47 *A 1939 photo*: Norma Barzman, *The Red and the Blacklist: The Intimate Memoir of a Hollywood Expatriate* (New York: Nation Books, 2003), 213.

47 *"a decidedly unconventional type"*: Letter from Bush to E. B. Wilson, December 15, 1938, Vannevar Bush Papers, Library of Congress.

48 *"Bush believed Shannon"*: Poundstone, *Fortune's Formula*, 21.

48 *brief notice*: "Youthful Instructor Wins Noble Award," *New York Times*, January 24, 1940.

48 *Back in Michigan*: "Institute Reports on Claude Shannon," *Otsego County Herald Times*, February 8, 1940.

48 *"I have a sneaking suspicion"*: Letter from Shannon to Bush, December 13, 1939, Shannon Papers.

49 *"queer algebra"*: Letter from Vannevar Bush to Barbara Burks, January 5, 1938, Bush Papers.

49 *"In these days"*: Quoted in Zachary, *Endless Frontier*, 70.

Chapter 6: *Cold Spring Harbor*

50 *"fittest families"*: Quoted in Garland E. Allen, "The Eugenics Record Office at Cold Spring Harbor: An Essay in Institutional History," *Osiris* 2, no. 2 (1986): 258.

50 *mailed state legislators*: Philip K. Wilson, "Harry Laughlin's Eugenic Crusade to Control the 'Socially Inadequate' in Progressive Era America," *Patterns of Prejudice* 36, no. 1 (2002): 49.

50 *a Nazi poster*: Brenda Jo Brueggeman, *Deaf Subjects: Between Identities and Places* (New York: New York University Press, 2009), 145.

51 *"biochemical deficiencies," etc.*: Allen, "The Eugenics Record Office," 239.

51 *"Sometimes a father"*: C. B. Davenport, *Naval Officers: Their Heredity and Development* (Washington, DC: Carnegie Institution of Washington, 1919), 29, quoted in Allen, "The Eugenics Record Office."

52 *"Thousands of stars"*: Frances Williston Burks, *Barbara's Philippine Journey* (Yonkers-on-Hudson, NY: World Book, 1921), 25.

53 *"surely Shannon is gifted"*: Letter from Burks to Bush, January 10, 1938, Bush Papers.

53 *"To advise a youth like Shannon"*: Quoted in Robert Price, "Oral History: Claude E. Shannon," IEEE Global History Network, July 28, 1982, www.ieeeghn.org/wiki/index.php/Oral-History:Claude_E._Shannon.

53 *"No work has been done"*: Shannon, "An Algebra for Theoretical Genetics," in *Claude Elwood Shannon: Collected Papers*, 920.

54 *"Although I looked"*: Letter from Shannon to Bush, February 16, 1939, Bush Papers.

54 *"Much of the power"*: Shannon, "An Algebra for Theoretical Genetics," in *Claude Elwood Shannon: Collected Papers*, 895.

54 *"genes are carried"*: Ibid., 892–93.

57 *"My theory has to do with"*: "Profile of Claude Shannon—Interview by Anthony Liversidge," in *Claude Elwood Shannon: Collected Papers*, xxvii.

58 *"he did not even know"*: Letter from Vannevar Bush to E. B. Wilson, December 15, 1938, Bush Papers.

58 *"very suitable"* . . . *"very much impressed"* . . . *"This, I feel strongly"*: Letter from Lowell J. Reed to Halbert L. Dunn, April 9, 1940, Bush Papers; letter from Dunn to Bush, April 19, 1940, Bush Papers; letter from Bush to Shannon, January 27, 1939, Bush Papers.

59 *"I had a good time"*: Shannon, interviewed by Hagemeyer, February 28, 1977.

59 *"few scientists"*: Letter from Burks to Bush, January 20, 1939, Bush Papers.

59 *"I doubt very much"*: Letter from Bush to Shannon, January 27, 1939, Bush Papers.

59 *"did not need to corrupt"*: Robert Gallager, personal communication, July 1, 2016.

60 *"After I had found the answers"* . . . "Too lazy": "Profile of Claude Shannon—Interview by Anthony Liversidge," xxviii, xxvii.

60 *ranked Shannon*: James F. Crow, "Shannon's Brief Foray into Genetics," *Genetics* 159, no. 3 (2001): 915–17.

60 *"that the work of all three"*: Crow, "Notes to 'An Algebra for Theoretical Genetics,'" in *Claude Elwood Shannon: Collected Papers*, 921.

60 *"Dear Dr. Bush"*: Letter from Shannon to Bush, February 16, 1939, Shannon Papers.

Chapter 7: *The Labs*

61 *"Real life mathematics"*: Bernard Beauzamy, "Real Life Mathematics," lecture, Dublin Mathematical Society, February 2001, scmsa.eu/archives/BB_real_life_maths_2001.htm.

61 *"unpredictable, irrational"*: Moulton-Barrett, *Graphotherapy*, 90.

62 *"Why don't you come out here"*: Norma Barzman, interviewed by the authors, December 21, 2014.

62 *"Christ-like"*: Quoted in Gertner, *The Idea Factory*, 121.

62 *"We spoke to each other"*: Barzman, *The Red and the Blacklist*, 378.

63 *"How can you"* . . . *"big trouble"* . . . *"was so loving"*: Norma Barzman, interviewed by the authors, December 21, 2014.

63 *denied them a room*: Poundstone, *Fortune's Formula*, 22.

63 *"I did not"*: Letter from Shannon to Bush, February 16, 1940, Bush Papers.

63 *"Well, I applied"*: Shannon, interviewed by Hagemeyer, February 28, 1977.

64 *"light-sensing reader system"*: Bradley O'Neill, "Dead Medium: The Comparator; the Rapid Selector," www.deadmedia.org/notes/1/017.html.

64 *"The only point I have in mind"*: Bush to Shannon, June 7, 1940. Vannevar Bush Papers, Library of Congress.

65 *"a very careful and formal person . . . rather frowned on"*: "Obituary: Thornton Carl Fry," American Astronomical Society, January 1, 1991.

65 *"Had I ever read"*: Thornton C. Fry, interviewed by Deirdre M. La Porte, Henry O. Pollak, and G. Baley Price, January 3–4, 1981, 4.

65 *"was where the future . . . the country's intellectual utopia . . . the crown jewel"*: Gertner, *The Idea Factory*, 1.

65 *"to consider what occurred"*: Ibid., 5.

66 *"the method of"*: "Improvement in Telegraphy," Patent Number US 174465 A

66 *"carry on scientific research"*: Walter Gifford, "The Prime Incentive," *Bell Laboratories Records,* vols. 1 and 2 (September 1925–September 1926), 18.

66 *"fundamental questions of physics"*: Gertner, *The Idea Factory*, 27.

67 *"When I first came"*: Henry Pollak, interviewed by the authors, August 7, 2014.

67 *"a fairyland company"*: Thornton C. Fry, interviewed by La Porte, Pollak, and Price, January 3–4, 1981, 10.

67 *"wraith-like and slow-moving . . . an almost spectral presence" . . . "he was allowed"*: Gertner, *The Idea Factory*, 28, 30.

68 *"I had the freedom to do anything I wanted"*: Shannon, interviewed by Albers, 1990.

69 *"Though the United States holds" . . . "The typical mathematician"*: Fry, "Industrial Mathematics," *Bell Systems Technical Journal* 20, no. 3 (July 1941): 256, 258.

70 *"pathetically ignorant of mathematics"*: Quoted in Gertner, *The Idea Factory*, 122.

70 *"Mathematicians are queer people"*: Thornton C. Fry, interviewed by La Porte, Pollak, and Price, January 3–4, 1981, 55.

70 *"our principle"*: Henry Pollak, interviewed by the authors, August 7, 2014.

70 *"there was nothing"*: Thornton C. Fry, interviewed by La Porte, Pollak, and Price, January 3–4, 1981, 56.

70 *"I was in the mathematics research group"*: Shannon, interviewed by Hagemeyer, February 28, 1977.

70 *exclusively by last names*: Thornton C. Fry, interviewed by La Porte, Pollak, and Price, January 3–4, 1981, 11.

71 *"There are a number of relays"*: Shannon, "A Theorem on Color Coding," Bell Laboratories, Memorandum 40-130-153, July 8, 1940.

72 *"The Use of the Lakatos-Hickman Relay"*: Shannon, "The Use of the Lakatos-Hickman Relay in a Subscriber Sender," Bell Laboratories, Memorandum 40-130-179, August 3, 1940.

73 *"I got quite a kick"*: Claude Shannon to Vannevar Bush. June 5, 1940. Vannevar Bush Papers, Library of Congress.

Chapter 8: *Princeton*

74 *"a man of extraordinary brilliancy"*: Norbert Wiener to J. R. Kline, April 10, 1941. Norbert Wiener Papers, MITA.

74 *"Mr. Shannon is one of the ablest graduates"*: H. B. Phillips cable to M. Morse, October 21, 1940.

75 *"He sees before him"*: Benedict Anderson, *Imagined Communities: Reflections on the Origin and Spread of Nationalism*, rev. ed. (New York: Verso, 2006), 57.

76 *"the smartest person"*: Letter from Shannon to William Aspray, October 27, 1987, Shannon Papers.

76 *"It is as if a wall"*: Hermann Weyl, *Space—Time—Matter*, 4th ed., trans. Henry L. Brose (New York: Dover, 1950), ix.

76 *"the fluctuations"*: Guizzo, "The Essential Message," 32.

77 *"I poured tea"*: Gertner, *The Idea Factory*, 121.

77 *"Your husband"*: Norma Barzman, interviewed by the authors, December 21, 2014.

78 *"The story is"*: Arthur Lewbel, "A Personal Tribute to Claude Shannon," www2.bc.edu/~lewbel/Shannon.html.

78 *"and he usually would walk along"*: Shannon, interviewed by Hagemeyer, February 28, 1977.

79 *"A kind of guilt or depression"*: Richard P. Feynman, *Surely You're Joking, Mr. Feynman*, reprint ed. (New York: Norton, 1997), 165.

79 *"Turn off the moon"*: Sam Coslow, "Turn Off the Moon," performed by Teddy Grace on *Turn on that Red Hot Heat*, rerelease, Hep Records, 1997.

79 *"You know where"*: Norma Barzman, interviewed by the authors, December 21, 2014.

80 *"America stands at the crossroads"*: Franklin Roosevelt, Proclamation 2425—Selective Service Registration, September 16, 1940.

81 *"Things were moving fast"*: Price, "Oral History: Claude E. Shannon."

81 *"if you can make yourself more useful"*: Shannon, interviewed by Hagemeyer, February 28, 1977.

81 *"I think he did the work"*: Maria Moulton-Barrett, interviewed by the authors, January 21, 2016.

81 *"There were those who protested"*: Bush, *Pieces of the Action*, 31–32.

Chapter 9: *Fire Control*

83 *"requirements of military training"*: Neva Reynolds. "Letter to Claude Shannon." February 10, 1941.

83 *"chopping wood"*: Mina Rees, "Warren Weaver," in National Academy of Sciences, *Biographical Memoirs*, vol. 57 (Washington, DC: National Academy Press, 1987), 494.

84 *"I didn't know"*: Warren Weaver, "Careers in Science," in *Listen to Leaders in Science*, ed. Albert Love and James Saxon Childers (Atlanta: Tupper & Love / David McKay, 1965), 276.

84 *"I think that God"*: Warren Weaver, *Science and Imagination: Selected Papers* (New York: Basic Books, 1967), 111.

84 *"I had a good capacity"*: Quoted in Rees, "Warren Weaver," 501.

85 *"Do not overestimate science"*: Warren Weaver, "Four Pieces of

Advice to Young People," in *The Project Physics Course Reader: Concepts of Motion*, ed. Gerald Holton et al. (New York: Holt, Rinehart & Winston, 1970), 22.

85 *epicure enough*: Thornton C. Fry, interviewed by La Porte, Pollak, and Price, January 3–4, 1981, 95.

86 *"At first thought"*: David A. Mindell, "Automation's Finest Hour: Bell Labs and Automatic Control in WWII," *IEEE Control Systems* 15 (1995): 72.

87 *"I found myself"*: Quoted in Howard Rheingold, *Tools for Thought* (Cambridge, MA: MIT Press, 2000), 103–4.

87 *"if my potentiometer"*: Quoted in Glenn Zorpette, "Parkinson's Gun Director," *IEEE Spectrum* 26, no. 4 (1989): 43.

88 *"I think England"*: Shannon, interviewed by Hagemeyer, February 28, 1977.

88 *"The wartime efforts"*: Mindell, "Automation's Finest Hour," 78.

89 *"special case of the transmission"*: R. B. Blackman, H. W. Bode, and C. E. Shannon, "Data Smoothing and Prediction in Fire-Control Systems," Summary Technical Report of Division 7, NDRC Vol. 1: Gunfire Control, ed. Harold Hazen (Washington, DC: Office of Scientific Research and Development, National Defense Research Committee, 1946).

89 *"He did some stunning work"*: Warren Weaver to Vannevar Bush, October 24, 1949. Bush Papers.

90 *"For a time"*: Letter from Warren Weaver to Vannevar Bush, October 24, 1949, Bush Papers.

Chapter 10: *A Six-Day Workweek*

91 *"This has not been"*: Vannevar Bush, "As We May Think," *Atlantic*, July 1945.

91 *"a warren of testing labs" . . . "a six-day workweek"*: Gertner, *The Idea Factory*, 26, 63.

91 *"It was a war"*: Fred Kaplan, "Scientists at War," *American Heritage* 34, no. 4 (June 1983): 49.

92 *"The attitude of many"*: J. Barkley Rosser, "Mathematics and

Mathematicians in World War II," in *A Century of Mathematics in America, Part 1*, ed. Peter Duren (Providence, RI: American Mathematical Society, 1988), 303.

93 *"He couldn't care less"* . . . *"He said he hated it"*: Maria Moulton-Barrett, interviewed by the authors, January 21, 2016.

94 *"those were busy times"*: Shannon, interviewed by Hagemeyer, February 28, 1977.

94 *"insisted to his dying day"*: Rosser, "Mathematics and Mathematicians," 304.

Chapter 11: *The Unspeakable System*

97 *"Such an image"*: Colin B. Burke, *It Wasn't All Magic: The Early Struggle to Automate Cryptanalysis, 1930s–1960s*. United States Cryptologic History, Special Series, Vol. 6, Center For Cryptologic History, National Security Agency, 2002.

97 *"Early in 1944"*: Warren F. Kimball, ed., *Churchill and Roosevelt: The Complete Correspondence*, vol. 3 (Princeton, NJ: Princeton University Press, 1984), 11.

98 *"some forty racks"*: Christopher H. Sterling, "Churchill and Intelligence—SIGSALY: Beginning the Digital Revolution," *Finest Hour* 149 (Winter 2010–11): 31.

98 *"rather like Rimsky-Korsakov's"*: David Kahn, *How I Discovered World War II's Greatest Spy and Other Stories of Intelligence and Code* (Boca Raton, FL: Auerbach, 2014), 147.

99 *"Accept distortion for security"*: Dave Tompkins, *How to Wreck a Nice Beach: The Vocoder from World War II to Hip-Hop, The Machine Speaks* (Chicago: Stop Smiling Books, 2011), 63.

99 *"Members working on the job"*: Andrew Hodges, *Alan Turing: The Enigma* (Princeton, NJ: Princeton University Press, 1983), 247.

99 *"It worked"*: Ibid., 312.

100 *"At a recent world fair"*: Bush, "As We May Think."

100 *"Phrt fdygui"*: Sterling, "Churchill and Intelligence," 34.

101 *"not a lot of laboratories"*: Shannon, interviewed by Hagemeyer, February 28, 1977.

101 *"a very down to earth discipline"*: Shannon, interviewed by Hagemeyer, February 28, 1977.

Chapter 12: *Turing*

104 *"Here [Turing] met a person"*: Hodges, *Alan Turing*, 314.

104 *"I think Turing had"* . . . *"We talked not at all"*: Price, "Oral History: Claude E. Shannon."

105 *"I reached New York"* . . . *"I had been intending"*: Alan Turing, "Alan Turing's Report from Washington DC, November 1942."

105 *"incomplete alliance"*: Andrew Hodges, "Alan Turing as UK-USA Link, 1942 Onwards," Alan Turing Internet Scrapbook, www.turing.org.uk/scrapbook/ukusa.html.

106 *"I am persuaded"*: Turing, "Alan Turing's Report from Washington DC, November 1942."

106 *"we would talk about"*: Price, "Oral History: Claude E. Shannon."

107 *"Well, back in '42"* . . . *"a very, very impressive guy"*: Shannon, interviewed by Hagemeyer, February 28, 1977.

108 *"While there we went over"*: Price, "Oral History: Claude E. Shannon."

Chapter 13: *Manhattan*

110 *"The ancient art of mathematics"*: Gareth Cook, "The Singular Mind of Terry Tao," *New York Times*, July 24, 2015.

110 *"There was a bedroom"* . . . *"He would find these common denominators"*: Maria Moulton-Barrett, interviewed by the authors, January 21, 2016.

111 *"Barney was an intellect"*: John Minck, "Inside HP: A Narrative History of Hewlett-Packard from 1939–1990," www.hpmemoryproject.org/timeline/john_minck/inside_hp_03.htm.

111 *"If the prospect of building devices"*: Thomas Perkins, *Valley Boy: The Education of Tom Perkins* (New York: Gotham, 2007), 72.

112 *"the world's first"*: Lawrence Fisher, "Bernard M. Oliver Is Dead at 79; Led Hewlett-Packard Research," *New York Times*, November 28, 1995.

112 *"We became friends"*: Arthur L. Norberg, "An Interview with Bernard More Oliver," Charles Babbage Institute for the History of Information Processing, August 9, 1985.

112 *"One day I was talking casually"*: John Pierce, "Creative Thinking," lecture, 1951.

113 *cowrote a key paper*: Bernard More Oliver, John Pierce, and Claude Shannon, "The Philosophy of PCM," *Proceedings of the IRE* 36, no. 11 (November 1948): 1324–31.

113 *"It turns out"*: Minck, "Inside HP."

113 *"I think it made him sick"*: Maria Moulton-Barrett, interviewed by the authors, January 21, 2016.

113–14 *"he had a certain type of impatience"*: Brockway McMillan, interviewed by the authors, January 4, 2016.

114 *"a very odd man"*: Gertner, *The Idea Factory*, 132.

114 *"He never argued"*: Ibid., 138.

114 *"genius is rarely able"*: George Henry Lewes, *The Principles of Success in Literature* (Berkeley: University of California Press, 1901), 98.

114 *"He was terribly, terribly secretive"*: Maria Moulton-Barrett, interviewed by the authors, January 21, 2016.

114 *"he was not someone"*: Robert Fano, interviewed by the authors, October 23, 2015.

114 *"He wouldn't have been"*: Brockway McMillan, interviewed by the authors, January 4, 2016.

114 *"My characterization of his smartness"*: Quoted in William Poundstone, *How to Predict the Unpredictable: The Art of Outsmarting Almost Anyone* (London: Oneworld, 2014).

114 *"He would have taken that"*: Peggy Shannon, interviewed by the authors, December 9, 2015.

115 *"These things sometimes"*: Shannon, interviewed by Hagemeyer, February 28, 1977.

115 *"He would go quiet"*: Maria Moulton-Barrett, interviewed by the authors, January 21, 2016.

Chapter 14: *The Utter Dark*

119 *"Repeat, please"*: Quoted in Samuel Carter, *Cyrus Field: Man of Two Worlds* (New York: Putnam, 1968), 167–68. See also Arthur C. Clarke, *Voice Across the Sea: The Story of Deep Sea Cable-Laying, 1858–1958* (London: Muller, 1958).

120 *"Down to the dark"*: Rudyard Kipling, "The Deep Sea Cables," in *Rudyard Kipling's Verse* (Garden City, NY: Doubleday, Page, 1922), quoted in Clarke, *Voice Across the Sea.*

120 *"nothing so much"*: "The Atlantic Telegraph Expedition," *Times* (London), July 15, 1858.

120 *"The very thought"*: Thomson, *The Life of William Thomson*, 1:362, quoted in Clarke, *Voice Across the Sea.*

121 *"is no longer"*: E. O. Wildman Whitehouse, "Report on a Series of Experimental Observations on Two Lengths of Submarine Electric Cable, Containing, in the Aggregate, 1,125 Miles of Wire, Being the Substance of a Paper Read Before the British Association for the Advancement of Science, at Glasgow, Sept. 14th, 1855," Brighton, 1855, 3, quoted in Bruce J. Hunt, "Scientists, Engineers, and Wildman Whitehouse: Measurement and Credibility in Early Cable Telegraphy," *British Journal for the History of Science* 29, no. 2 (1996): 158.

122 *"a fiction"*: E. O. Wildman Whitehouse, "The Law of Squares—Is It Applicable or Not to the Transmission of Signals in Submarine Circuits?," *Athenaeum,* August 30, 1856, 1092–93, quoted in Hunt, "Scientists, Engineers, and Wildman Whitehouse."

122 *"fallacious"*: Quoted in Thomson, *The Life of William Thomson,* 1:330.

122 *"The further the electricity"*: Donard De Cogan, "Dr E.O.W. Whitehouse and the 1858 Trans-Atlantic Cable," *History of Technology* 10 (1985): 2.

122 *"Forty-eight words"*: *Report of the Joint Committee to Inquire into the Construction of Submarine Telegraph Cables* (London: Eyre & Spottiswoode, 1861), 237.

123 *"bullied"*: Clarke, *Voice Across the Sea.*

Chapter 15: *From Intelligence to Information*

125 *most complex machines*: See Mindell, *Between Human and Machine*, 107ff.

126 *came to understand heat*: See Matthew Crawford, *Shop Class as Soulcraft: An Inquiry into the Value of Work* (New York: Penguin, 2010), 22–23.

127 *a bandwidth of 3,000 hertz*: Guizzo, "The Essential Message," 26.

128 *between continuous signals*: See ibid., 27.

128 *how to send telegraph and telephone signals*: It was already "common practice to send telegraph and telephone signals on the same wires," but Nyquist's work, by minimizing interference from telegraph signals, led to clearer telephone calls. See John Pierce, *An Introduction to Information Theory: Symbols, Signals and Noise*, 2nd ed. (New York: Dover, 1980), 38.

128 *"the world of technical communications"*: James L. Massey, "Information Theory: The Copernican System of Communications," *IEEE Communications Magazine* 22, no. 12 (1984): 27, cited in Guizzo, "The Essential Message."

128 *"by the speed of transmission"*: Harry Nyquist, "Certain Factors Affecting Telegraph Speed," *Bell System Technical Journal*, April 1924, 332. See also Nyquist, "Certain Topics in Telegraph Transmission Theory," *Transactions of the AIEE* 47 (April 1928): 617–44.

130 *the tantalizing suggestion*: On the relationship between Nyquist's and Shannon's work, see William Aspray, "The Scientific Conceptualization of Information: A Survey," *IEEE Annals of the History of Computing* 7, no. 2 (1985): 121.

130 *"an important influence"*: Robert Price, "A Conversation with Claude Shannon: One Man's Approach to Problem Solving," *IEEE Communications Magazine* 22, no. 6 (May 1984): 123.

131 *"a ball rolling," etc.*: Ralph Hartley, "Transmission of Information," *Bell System Technical Journal* 7, no. 3 (July 1928): 536–38.

134 *a 20-letter telegram*: Guizzo, "The Essential Message," 25.

134 *"practical engineering value"*: Hartley, "Transmission of Information," 539.

136 *"within the realm"*: Ibid., 563.

136 *"very bright in some ways"*: Shannon, interviewed by Hagemeyer, February 28, 1977.

137 *"appears to have taken"*: Pierce, *An Introduction to Information Theory*, 40.

137 *"It came as a bomb"*: Pierce, "The Early Days of Information Theory," 4.

Chapter 16: *The Bomb*

138 *"The fundamental problem"*: Claude Shannon, "A Mathematical Theory of Communication," in *Claude Elwood Shannon: Collected Papers*, 5.

138 "selected from a set": Ibid. Emphasis in original.

139 diagram: Ibid., 7.

140 *genes as information-bearers*: Around the time of the publication of "A Mathematical Theory," Shannon attempted to estimate the amount of information, in bits, in the human genome. Shannon, untitled document, July 12, 1949, Shannon Papers; see Gleick, *The Information*, 231.

141 *"what would be the simplest source"*: Price, "Oral History: Claude E. Shannon."

141 *A fair coin*: For simplicity's sake, this section deals only with discrete symbols, rather than continuous ones.

141 *One bit is the amount of information*: We discuss Shannon's revolutionary response to the problem of noise below. Here we are leaving noise out of consideration; but one should bear in mind that if we were to learn about the result of a coin toss through a noisy channel, we may still get less than one bit of information, even if the coin toss itself were fair.

141 *"a device with two stable positions"*: Shannon, "Mathematical Theory," 6.

143 diagram: Ibid., 20.

144 *"average surprise"*: In other words, if the coin is weighted so that the probability of tails is low, then the larger surprise of seeing tails is balanced against the longer odds of seeing that outcome.

144 *"I don't regard it"*: Price, "A Conversation with Claude Shannon," 123.

145 *three basic Morse characters*: This slightly simplifies Shannon's example, which allowed for two kinds of spaces, those following letters and those following words.

146 *"XFOML," etc.*: Shannon, "Mathematical Theory," 14.

148 *"The particular sequence"*: Ibid., 15.

148 *unpublished spoof*: Shannon, July 4, 1949, Shannon Papers.

150 *"Now, in English"*: Poe, "The Gold-Bug," 101–2.

150 *switching code alphabets*: Examples from David Kahn, *The Codebreakers: The Story of Secret Writing* (New York: Macmillan, 1953), 749.

151 *"I wrote"*: Price, "Oral History: Claude E. Shannon."

151 *"Roughly, redundancy means"*: Ibid. 744.

151 *"MST PPL"*: Shannon, "Information Theory," in *Encyclopaedia Britannica*, 14th ed., reprinted in *Claude Elwood Shannon: Collected Papers*, 216.

152 *"When we write English"*: Shannon, "Mathematical Theory," 25.

152 *"certain known results"*: Ibid. This comment aside, cryptography is not an explicit topic of Shannon's 1948 paper. But because Shannon observed that his work on cryptography and information theory were mutually influential, we have discussed the overlap between the two fields in this chapter, especially to illustrate the importance to both of the concept of redundancy.

153 *"A S-M-A-L-L"*: Gleick, *The Information*, 230.

154 *Shannon explained*: Shannon, "Information Theory," 216.

157 *like transmitting power*: Massey, "Information Theory," 27.

159 *Kahn illustrates this point*: Kahn, *The Codebreakers*, 747.

160 *a code like this*: This example is cited in Guizzo, "The Essential Message," 40.

160 *combining the advantages of codes*: In Shannon's more precise terms, combining "source coding" and "channel coding."

161 *"Up until that time"*: Robert Gallager, interviewed by the authors, August 8, 2014.

161 *founded a new field and solved most of its problems*: David J. C. MacKay, *Information Theory, Inference, and Learning Algorithms* (Cambridge: Cambridge University Press, 2003), 14.

162 *information reduces "entropy"*: Specifically, in Shannon's terms, entropy can be thought of as uncertainty, and information can be thought of as the amount of uncertainty that is reduced by an observation, measurement, or description.

162 *"And more importantly"*: An early version of the anecdote appears in Myron Tribus and Edward C. McIrving, "Energy and Information," *Scientific American* 225 (1971): 179–88.

162 *Hungarian physicist Leo Szilard*: For a more complete story of Szilard and "Maxwell's Demon," see Gleick, *The Information*, 275–80, and George Johnson, *Fire in the Mind* (New York: Vintage, 1995), 114–21.

163 *"Organisms organize"*: Gleick, *The Information*, 281.

Chapter 17: *Building a Bandwagon*

165 *"The Magna Carta" . . . "Without Claude's work" . . . "A major contribution" . . . "A universal clue" . . . "I reread it every year" . . . "I know of no greater"*: Horgan, "Claude E. Shannon," 22A; Lewbel, "A Personal Tribute to Claude Shannon"; Roy Rosenzweig Center for History and New Media, George Mason University, "Remembering Claude Shannon," March–August 2001, chnm.gmu.edu/digitalhistory/links/cached/chapter6/6_19b_surveyresponse.htm; Robert W. Lucky, *Silicon Dreams*, quoted in Lee Dembart, "Book Review: Putting on Thinking Caps Over Artificial Intelligence," *Los Angeles Times*, August 15, 1989.

166 *"While, of course"*: R. J. McEliece, *The Theory of Information and Coding: Student Edition* (New York: Cambridge University Press, 2004), 13.

167 *"Weaver became the expositor"*: Ronald R. Kline, *The Cybernetics Moment: Or Why We Call Our Age the Information Age* (Baltimore: Johns Hopkins University Press, 2015), 122.

167 *"Radar won the war"*: Quoted in Wolfgang Saxon, "Albert G. Hill, 86, Who Helped Develop Radar In World War II," *New York Times*, October 29, 1996.

168 *"press's new series"*: Letter from Louis Ridenour to Warren Weaver, March 21, 1949; letter from Ridenour to Mervin Kelly,

April 12, 1949, Institute of Communications Research, Record Series 13/5/1, University of Illinois Archives.

168 *bestselling*: Jorge Reina Schement and Brent D. Ruben, *Between Communication and Information* 4 (New Brunswick, NJ: Transaction, 1993), 53.

169 *"There could very easily"*: Letter from Weaver to Ridenour, November 17, 1949, Shannon Papers, quoted in Kline.

Chapter 18: *Mathematical Intentions, Honorable and Otherwise*

170 *"I have read your book"*: Adam Sedgwick, "Letter to Charles Darwin," November 24, 1859.

170 *"seemed initially too simple"*: Sylvia Nasar, *A Beautiful Mind: The Life of Mathematical Genius and Nobel Laureate John Nash* (New York: Simon & Schuster, 1998).

171 *"I have always wanted"*: J. L. Doob, "Review of *A Mathematical Theory of Communication,*" *Mathematical Review* 10 (1949): 133.

171 *"Robert was all for it"*: Naresh Jain, "Record of the Celebration of the Life of Joseph Leo Doob," www.math.uiuc.edu/People/doob_record.html.

171 *"Are there infinitely many"*: Lashi Bandara, "Explainer: The Point of Pure Mathematics," *The Conversation*, August 1, 2011, theconversation.com/explainer-the-point-of-pure-mathematics-2385.

172 *"must learn the art of numbers"*: Uta C. Merzbach and Carl B. Boyer, *A History of Mathematics*, 3rd ed. (Hoboken, NJ: John Wiley & Sons, 2011), 77.

172 *"There is a tale told"*: Ibid., 91.

172 *"manifesto for mathematics"* . . . *"Beauty is the first test"*: G. H. Hardy, *A Mathematician's Apology* (Cambridge: Cambridge University Press, 2013), back matter, 85, 135.

172 *"The discussion is suggestive"*: Doob, "Review of *A Mathematical Theory of Communication.*"

173 "*LIVERSIDGE: When* The Mathematical Theory": "Profile of Claude Shannon—Interview by Anthony Liversidge," in *Claude Elwood Shannon: Collected Papers*, xxvii.

173 *"the occasional liberties taken"*: Shannon, "Mathematical Theory," 50.

173 *"When Shannon's paper appeared"*: Solomon W. Golomb, "Claude Elwood Shannon," *Notices of the AMS* 49, no. 1 (2001): 9.

174 *"Distinguished and accomplished as Doob was"*: Edward O. Thorp, personal communication, April 8, 2017.

174 *"It turned out that everything he claimed"*: Sergio Verdú, "Fireside Chat on the Life of Claude Shannon," www.youtube.com/watch?v=YEt9P2kp9BE.

Chapter 19: *Wiener*

175 *"the American John Von Neumann"*: Nasar, *A Beautiful Mind,* 135.

175 *"I had full liberty . . . He would begin the discussion"*: Norbert Wiener, *Ex-Prodigy: My Childhood and Youth* (Cambridge, MA: MIT Press, 1964), 67–68.

176 *"From every angle"*: Paul Samuelson, "Some Memories of Norbert Wiener," in *The Legacy of Norbert Wiener: A Centennial Symposium* (Cambridge, MA: American Mathematical Society, 1994).

176 *"In appearance and behaviour"*: Hans Freudenthal, "Norbert Wiener," in *Complete Dictionary of Scientific Biography*, www.encyclopedia.com/people/science-and-technology/mathematics-biographies/norbert-wiener.

177 *"Can you show me where"*: Samuelson, "Some Memories of Norbert Wiener."

177 *Shannon had taken Wiener's class*: Price, "Oral History: Claude E. Shannon."

177 *"an idol of mine"*: "Profile of Claude Shannon—Interview by Anthony Liversidge," in *Claude Elwood Shannon: Collected Papers*, xxxii.

177 *"Shannon and I"*: Norbert Wiener, *I Am a Mathematician,* 179.

177 *"Under these circumstances"*: Norbert Wiener to Walter Pitts, April 4, 1947. Norbert Wiener Papers, MITA.

178 *"total irresponsibleness"*: Norbert Wiener to Arturo Rosenblueth, April 16, 1947. Norbert Wiener Papers, MITA.

178 *"lost priority"*: Norbert Wiener to Warren McCulloch, April 5, 1947. Norbert Wiener Papers, MITA.

178 *"One of my competitors"*: Norbert Wiener to Arturo Rosenblueth, April 16, 1947. Norbert Wiener Papers, MITA.

178 *"The Bell people"*: Norbert Wiener to Warren McCulloch, May 2, 1927, Norbert Wiener Papers, MITA.

178 *"the entire field"*: Norbert Wiener, *Cybernetics, or Control and Communication in the Animal and the Machine*, 2nd ed. (Cambridge, MA: MIT Press, 1961), 11.

178 *"might be comparable"*: John Platt, "Books That Make a Year's Reading and a Lifetime's Enrichment," *New York Times*, February 2, 1964.

178 *"the biggest bite"*: Gregory Bateson, *Steps to an Ecology of the Mind*, 484.

178 *"Wiener in a sense"*: Thomas Kailath, interviewed by the authors, June 2, 2016.

179 *"in fact, there is no evidence"*: Sergio Verdú, interviewed by the authors, September 6, 2015.

180 *"When I talked to Norbert"*: Price, "Oral History: Claude E. Shannon."

180 *"I don't think Wiener"*: Claude Shannon, in *Claude Elwood Shannon: Collected Papers*, xix

Chapter 20: *A Transformative Year*

181 *"For most people"*: Nasar, *A Beautiful Mind*, 228.

182 *"My mother was eternally grateful"*: Peggy Shannon, interviewed by the authors, December 9, 2015.

183 *"students and faculty"*: "Who We Are," Douglass Residential College, Rutgers University, douglass.rutgers.edu/history.

183 *"fortunately had good grades"*: Betty Shannon, interviewed by the authors, November 12, 2015.

183 *"the best offer"*: Betty Shannon, interviewed by the authors, November 12, 2015.

184 *"he was very quiet"*: Ibid.

184 *"not very formal"*: Ibid.

184 *"I need my wife"*: Quoted in Monique Frize, Peter Frize, and Nadine Faulkner, *The Bold and the Brave* (Ottawa, Canada: University of Ottawa Press, 2009), 285.

184 *"I think I'm more visual than symbolic"*: Shannon, interviewed by Albers, 1990.

184 *"He didn't know math" . . . "He had a weird insight"*: Quoted in Kevin Coughlin, "Claude Shannon: The Genius of the Digital Age," *Star-Ledger* (New Jersey), February 28, 2001.

185 *"wouldn't go out of his way"*: Quoted in Eugene Chiu et al., "Mathematical Theory of Claude Shannon," December 2001, web.mit.edu/6.933/www/Fall2001/Shannon1.pdf.

185 *"some of his early papers"*: Betty Shannon, interviewed by the authors, November 12, 2015.

Chapter 21: *TMI*

186 *"Great scientific theories"*: Francis Bello, "The Information Theory," *Fortune*, December 1953, 136–58.

187 *"Much as I wish"*: Quoted in Kline, *The Cybernetics Moment*, 124.

187 *"It may be no exaggeration"*: Bello, "The Information Theory," 136.

187 *"Gaylord native son"*: "Gaylord's Claude Shannon: 'Einstein of Mathematical Theory,'" *Gaylord Herald Times*, October 11, 2000.

187 *"There were many"*: Poundstone, *Fortune's Formula*, 15.

188 *"What kind of man"*: Bello, "The Young Scientists," *Fortune*, June 1954, 142.

188 "OMNI: *Did you feel*": "Profile of Claude Shannon—Interview by Anthony Liversidge," in *Claude Elwood Shannon: Collected Papers*, xxviii.

189 *"The expansion of the applications"*: L. A. de Rosa, "In Which Fields Do We Graze?," *IRE Transactions on Information Theory* 1, no. 3 (1955): 2.

189 *"Information theory has," etc.*: Shannon, "The Bandwagon," 3.

191 *"Claude Shannon was"*: Robert G. Gallager, "Claude E. Shannon: A Retrospective on His Life, Work, and Impact," *IEEE Transactions on Information Theory* 47, no. 7 (2001): 2694.

192 *"He got a little irritated"*: Quoted in Omar Aftab et al., "Information Theory and the Digital Age," 10, web.mit.edu/6.933/www/Fall2001/Shannon2.pdf.

192 *"I didn't like the term"*: Quoted in ibid., 9.

Chapter 22: *"We Urgently Need the Assistance of Dr. Claude E. Shannon"*

193 *"Dear Dr. Kelly"*: Letter from Walter B. Smith to M. J. Kelly, May 4, 1951, National Security Agency, www.nsa.gov/public_info/_files/friedmanDocuments/PanelCommitteeandBoardRecords/FOLDER_393/41745239078444.pdf.

194 *"I hope very much"*: Letter from Kingman Douglass to J. N. Wenger, May 7, 1951, National Security Agency, www.nsa.gov/public_info/_files/friedmanDocuments/PanelCommitteeandBoardRecords/FOLDER_393/41745239078444.pdf.

194 *"One of the first"* . . . *"message externals"*: David A. Hatch and Robert Louis Benson, "The Korean War: The SIGINT Background," National Security Agency, www.nsa.gov/public_info/declass/korean_war/sigint_bg.shtml.

194 *"I spoke to Shannon today"*: Handwritten note on top of letter from Kingman Douglass to J. N. Wenger, May 7, 1951, National Security Agency, www.nsa.gov/public_info/_files/friedmanDocuments/PanelCommitteeandBoardRecords/FOLDER_393/41745239078444.pdf.

195 *"While there have been"*: Letter from Kelly to Smith, National Security Agency, www.nsa.gov/public_info/_files/friedmanDocuments/PanelCommitteeandBoardRecords/FOLDER_393/41745139078434.pdf.

195 *"to be plucked"* . . . *"how to keep"*: Nasar, *A Beautiful Mind*, 107, 106.

196 *"the fundamental purpose"*: National Security Agency, "NSA Regulation Number 11-3," January 22, 1953, ia601409.us.archive.org/16/items/41788579082758/41788579082758.pdf.

196 *"Because a considerable portion"*: Anne S. Brown, "Historical Study: The National Security Agency Scientific Advisory Board, 1952–1963" (Washington, DC: NSA Historian, Office of Central Reference, 1965), 4.

197 *"Lacking accessible, secure areas"*: Ibid.

197 *"Price: And you were on the board"*: Price, "Oral History: Claude E. Shannon."

Chapter 23: *The Man-Machines*

199 *"Could a machine think?"*: Ludwig Wittgenstein, *Philosophical Investigations*, trans. G. E. M. Anscombe et al., ed. P. M. S. Hacker and Joachim Schulte, 4th ed. (Malden, MA: Blackwell, 2009), 359–60.

199 *"I'm a machine"*: Quoted in Horgan, "Claude E. Shannon," 22B.

199 *"had earned the right"*: Henry Pollak, interviewed by the authors, August 7, 2014.

200 *"it seemed lost on Shannon"*: Gertner, *The Idea Factory*, 141.

201 *"If you read Science Fiction"*: Letter from Shannon to Warren S. McCulloch, August 23, 1949, Warren S. McCulloch Papers, American Philosophical Society, Series I, Shannon correspondence.

201 *"to this day"*: Poundstone, *Fortune's Formula*, 61.

201 *"Dear Dr. Shannon"* . . . *"to complete and verify"* . . . *"Dear Sir"*: Letter from Earle L. Morrow to Shannon, December 26, 1957, Shannon Papers; letter from George C. Paro to Shannon, November 17, 1960, Shannon Papers; letter from Daniel J. Quinlan to Shannon, April 13, 1953, Shannon Papers.

202 *"I think the history of science"*: Shannon, "Development of Communication and Computing, and My Hobby," lecture, Inamori Foundation, Kyoto, Japan, November 1985, www.kyotoprize.org/wp/wp-content/uploads/2016/02/1kB_lct_EN.pdf.

202 *"with the possible capabilities"*: [AU: Please supply reference.]

203 *"I went out and got him"* . . . *"Giving it to a grown man"*: "Profile of Claude Shannon—Interview by Anthony Liversidge," in *Claude Elwood Shannon: Collected Papers*, xxii.

203 *"We did all this"*: Quoted in Timothy Johnson, "Claude Elwood Shannon: Information Theorist," Shannon Papers.

204 *"Hello"*: Bell Labs, "Claude Shannon Demonstrates Machine Learning," www.youtube.com/watch?v=vPKkXibQXGA.

205 *"I was told that"*: Henry Pollak, interviewed by the authors, August 7, 2014.

205 *"Mouse with a Memory"*: *Time*, May 19, 1952, 59.

206 *"It is all too human"*: "Presentation of a Maze-Solving Machine," in *Cybernetics: Transactions of the Eighth Conference March 15–16,*

1951, ed. Heinz von Foerster, Margaret Mead, and Hans Lukas Teuber (New York: Josiah Macy, Jr. Foundation, 1952), 179.

206 *"The fascination of watching Shannon's innocent rat"*: "Note by the Editors," in ibid., xvii.

206 *"a demonstration device"*: Letter from Shannon to Irene Angus, August 8, 1952, Shannon Papers, cited in Gertner, *The Idea Factory*.

207 *"The design of game playing machines"*: Shannon, "Game Playing Machines," in *Claude Elwood Shannon: Collected Papers*, 786.

207 *"My fondest dream"*: Letter from Shannon to Angus, August 8, 1952, Shannon Papers.

207 *"In the long run"*: Brock Brewer, "The Man-Machines May Talk First to Dr. Shannon," *Vogue*, April 15, 1963, 139.

207 *"The Man-Machines" . . . "Dr. Claude E. Shannon"*: Ibid., 89.

208 *"you have to think of problems"*: Ibid.

208 *"First, how can we give computers"*: Ibid., 139.

208 *"I believe that today"*: Shannon, interviewed by Hagemeyer, February 28, 1977.

209 *"thinking is sort of the last thing"*: Shannon, interviewed by Albers, 1990.

209 *"We artificial intelligence people"*: Shannon, untitled document, 1984, Shannon Papers.

Chapter 24: *The Game of Kings*

210 *"an oriental sorcerer"*: Quoted in Tom Standage, *The Turk: The Life and Times of the Famous Eighteenth-Century Chess-Playing Machine* (New York: Walker, 2002), 23.

210 *"the Turk—even he"*: Silas Weir Mitchell, "The Last of a Veteran Chess Player." *Chess Monthly*, 1857.

211 *"who is never to be seen"*: Poe, "Maelzel's Chess Player," in *The Complete Tales of Edgar Allan Poe* (New York: Vintage Books, 1975), 438.

211 *"at least one supervisor"*: Horgan, "Claude E. Shannon," 22A.

211 *"Most of us"*: Brockway McMillan, interviewed by the authors, January 4, 2016.

212 *"Botvinnik was worried"*: "Profile of Claude Shannon—Interview by Anthony Liversidge," in *Claude Elwood Shannon: Collected Papers*, xxix.

212 *Another incident*: Peggy Shannon, interviewed by the authors, December 9, 2015.

212 *"There have been few new ideas"*: Norman Whaland, "A Computer Chess Tutorial," *Byte*, October 1978, 168.

212 *"Although perhaps of no practical importance," etc.*: "Programming a Computer for Playing Chess," in *Claude Elwood Shannon: Collected Papers*, 637–38, 650–54.

215 *"Claude went hog wild"*: "Profile of Claude Shannon—Interview by Anthony Liversidge," in *Claude Elwood Shannon: Collected Papers*, xxxi.

215 *"From a behavioristic point of view"*: Shannon, "A Chess-Playing Machine," in *Claude Elwood Shannon: Collected Papers*, 655.

216 *"just foolish logic"*: H. J. van den Herik, "An Interview with Claude Shannon (September 25, 1980 in Linz, Austria)," *ICCA Journal* 12, no. 4 (1989): 225.

Chapter 25: *Constructive Dissatisfaction*

217 *"A very small percentage," etc.*: Claude Shannon, "Creative Thinking," March 20, 1952, in *Claude Shannon's Miscellaneous Writings*, ed. N. J. A. Sloane and Aaron D. Wyner (Murray Hill, NJ: Mathematical Sciences Research Center, AT&T Bell Laboratories, 1993), 528–39.

Chapter 26: *Professor Shannon*

223 *"There is an active structure"*: Quoted in Gertner, *The Idea Factory*, 146.

223 *"I am having a very enjoyable time"*: Quoted in Poundstone, *Fortune's Formula*, 27.

224 *"From the questions"*: Letter from Shannon to John Riordan, February 20, 1956, Shannon Papers.

224 *"In case men's lives"*: Shannon, Assorted lecture notes, n.d., Shannon Papers.

225 *"The following analysis"*: Shannon, "The Portfolio Problem," n.d., Shannon Papers.

225 *"It always seemed to me"*: Letter from Shannon to H. W. Bode, October 3, 1956, Shannon Papers, quoted in Gertner, *The Idea Factory*, 146.

226 *"Having spent fifteen years"*: Quoted in Poundstone, *Fortune's Formula*, 27.

226 *"Shannon is one of the great people"*: Henry Pollak, interviewed by the authors, August 7, 2014.

227 *said to have wondered aloud*: Thomas Kailath, interviewed by the authors, June 2, 2016.

227 *"You are going to God's country"*: "Remembering Claude Shannon," chnm.gmu.edu/digitalhistory/links/cached/chapter6/6_19b_surveyresponse.htm.

227 *"a three-sided verandah"*: National Register of Historic Places application, Edmund Dwight House, Massachusetts Cultural Resource Information System, mhc-macris.net/Details.aspx?MhcId=WNT.19.

228 *"Although he continued"*: Robert E. Kahn, "A Tribute to Claude E. Shannon," *IEEE Communications Magazine*, July 2001, 18.

228 *unicycles, in every permutation*: Ronald Graham, interviewed by the authors, August, 23, 2014.

229 *"I've always pursued my interests"*: Quoted in Horgan, "Claude E. Shannon," 22A.

229 *"These were things"*: Robert Gallager, interviewed by the authors, August 8, 2014.

229 *"He was really lionized"*: Quoted in Chiu et al., "Mathematical Theory of Claude Shannon."

229 *"If I'm going to spend"*: Leonard Kleinrock, interviewed by the authors, September 16, 2016.

229 *"I can't be an advisor"*: Quoted in Guizzo, "The Essential Message," 61.

230 *"I was in such awe of him"*: Robert Gallager, interviewed by the authors, August 8, 2014.

230 *"I always felt honored"*: Leonard Kleinrock, interviewed by the authors, September 16, 2016.

230 *"I was just so impressed"*: Quoted in Guizzo, "The Essential Message," 59.

230 *"His classes were like"*: Leonard Kleinrock, interviewed by the authors, September 16, 2016.

231 *"For some problems"*: Quoted in ibid., 60.

231 *"He was not the sort of person"*: Robert Gallager, interviewed by the authors, August 8, 2014.

231 *"We all revered Shannon"*: Quoted in Guizzo, "The Essential Message," 59.

231 *"He said, 'Why don't you'"*: Leonard Kleinrock, interviewed by the authors, September 26, 2016.

232 *"Shannon's favorite thing to do"*: Larry Roberts, interviewed by the authors, September 16, 2016.

232 *"I had what I thought"*: Robert Gallager, interviewed by the authors, August 8, 2014.

233 *"People would go in"*: Irwin Jacobs, interviewed by the authors, January 1, 2015.

233 *"He did a lot of work at home"* . . . *"Did you know"*: Peggy Shannon, interviewed by the authors, December 9, 2015.

235 *"golden age of information theory"*: Thomas Kailath, interviewed by the authors, June 2, 2016.

235 *"The intellectual content"*: Anthony Ephremides, interviewed by the authors, May 31, 2016.

235 *"I believe that scientists"*: "Profile of Claude Shannon—Interview by Anthony Liversidge," in *Claude Elwood Shannon: Collected Papers*, xxiii.

235 *"no mathematician should ever"*: Hardy, *A Mathematician's Apology*, 70.

235 *"I started telling him about it"*: Henry Pollak, interviewed by the authors, August 7, 2014.

236 *"For everybody who built"*: Robert Gallager, interviewed by the authors, August 8, 2014.

236 *"I think that this present century"*: Quoted in Gertner, *The Idea Factory*, 317.

Chapter 27: *Inside Information*

238 *"Much of the conversation"*: Peggy Shannon, interviewed by the authors, December 9, 2015.

239 *"I had a good opinion"*: "Profile of Claude Shannon—Interview by Anthony Liversidge," in *Claude Elwood Shannon: Collected Papers*, xxvii.

240 *"I run the checkbook"*: Betty Shannon, interviewed by Albers, 1990.

240 *"their work in the stock market"*: Peggy Shannon, interviewed by the authors, December 9, 2015.

240 *"Nah, the commissions would kill you"*: Quoted in Poundstone, *Fortune's Formula*, 208.

240 *"I even did some work"*: "Profile of Claude Shannon—Interview by Anthony Liversidge," in *Claude Elwood Shannon: Collected Papers*, xxiv–xxv.

242 *"used common sense"*: Peggy Shannon, interviewed by the authors, December 9, 2015.

242 *"I make my money"*: Price, "Claude E. Shannon: An Interview."

242 *"Inside information"*: Poundstone, *Fortune's Formula*, 21.

Chapter 28: *A Gadgeteer's Paradise*

244 *"The secretary warned me," etc.*: Edward O. Thorp, "The Invention of the First Wearable Computer," *Proceedings of the 2nd IEEE International Symposium on Wearable Computers*, October 1998, 4–8.

246 *"The division of labor"*: Thorp, personal communication, March 24, 2017.

Chapter 29: *Peculiar Motions*

247 *"Do you mind"*: Lewbel, "A Personal Tribute to Claude Shannon."

248 *"As a physical experiment" . . . "He just showed up"*: Arthur Lewbel, interviewed by the authors, August 8, 2014.

248 *"One nice thing"*: Lewbel, "A Personal Tribute to Claude Shannon."

249 *"The Juggling Club"*: Peggy Shannon, interviewed by the authors, December 9, 2015.

249 *"When Galileo wanted"*: Ronald Graham, interviewed by the authors, August, 23, 2014.

249 *"it was something"*: Gertner, *The Idea Factory*, 319–20.

250 *"mathematics is often described"*: Persi Diaconis and Ron Graham, *Magical Mathematics* (Princeton, NJ: Princeton University Press, 2012), 137.

250 *"next time you see some jugglers"*: Burkard Polster, "The Mathematics of Juggling," 1, qedcat.com/articles/juggling_survey.pdf.

250 *"He liked peculiar motions"*: Arthur Lewbel, interviewed by the authors, August 8, 2014.

250 *"is complex enough"*: Peter J. Beek and Arthur Lewbel, "The Science of Juggling," *Scientific American* 273, no. 5 (November 1995): 92.

252 *"'Do you think juggling's a mere trick,'" etc.*: Shannon, "Scientific Aspects of Juggling," in *Claude Elwood Shannon: Collected Papers*, 850–57.

254 *"In his twenty years' devotion"*: "Enrico Rastelli," *Vanity Fair*, February 1932, 49.

256 *"It all started"*: Shannon, "Claude Shannon's No Drop Juggling Diorama," in *Claude Elwood Shannon: Collected Papers*, 847.

256 *"each side making a catch"*: Beek and Lewbel, "The Science of Juggling," 97.

256 *"The greatest numbers jugglers"*: Shannon, "Claude Shannon's No Drop Juggling Diorama," 849.

Chapter 30: *Kyoto*

257 *"He was a very modest guy"*: Betty Shannon, interviewed by the authors, November 12, 2015.

257 *"I don't think I was ever motivated"*: "Profile of Claude Shannon—Interview by Anthony Liversidge," in *Claude Elwood Shannon: Collected Papers*, xxiv.

258 *"the calls would come"*: Peggy Shannon, interviewed by the authors, December 9, 2015.

258 *"brilliant contributions"*: National Medal of Science citation, www.nationalmedals.org/laureates/claude-e-shannon.

258 *"eleven men whose lifelong purpose"*: Lyndon Johnson, Remarks Upon Presenting the National Medal of Science Awards for 1966, February 6, 1967.

258 *"I was seven"*: Peggy Shannon, interviewed by the authors, December 9, 2015.

259 *"to get something out of Claude"*: Letter from Rud: Kompfner to Pierce, June 1, 1977, quoted in Gertner, *The Idea Factory,* 323.

259 *"An American driving in England"*: Shannon, "The Fourth-Dimensional Twist, or a Modest Proposal in Aid of the American Driver in England," 1978, Shannon Papers.

261 *"he'd get an award"*: Peggy Shannon, interviewed by the authors, December 9, 2015.

261 *"I have never seen such stage fright"*: Quoted in University of California Television, "Claude Shannon: Father of the Information Age," 2002, www.youtube.com/watch?v=z2Whj_nL-x8.

261 *"He just felt that people"*: Quoted in ibid.

261 *"Since our retirement"*: Letter from Shannon to Ben [last name unknown], November 15, 1980, Shannon Papers.

262 *"we always thought that information theory"*: Thomas Kailath, interviewed by the authors, June 2, 2016.

262 *"he didn't like to boast"*: Arthur Lewbel, interviewed by the authors, August 8, 2014.

262 *nominated for the Nobel in physics*: Nomination Database, NobelPrize.org, www.nobelprize.org/nomination/archive/show_people.php?id=10947.

263 *"You know, there's no Nobel"*: Quoted in Flo Conway and Jim Siegelman, *Dark Hero of the Information Age: In Search of Norbert Wiener, the Father of Cybernetics* (New York: Basic Books, 2005), 394, n. 327.

263 *"After a quarter of a century"*: Kazuo Inamori, "Philosophy," Inamori Foundation, April 12, 1984, www.inamori-f.or.jp/en/kyoto_prize/.

264 *"The Kyoto prize"*: "Kyoto Prize 2015: Inamori Foundation Announces This Year's Laureates," June 19, 2015, goo.gl/kYNzdJ.

265 *"two strong women"*: Peggy Shannon, interviewed by the authors, December 9, 2015.

265 *"I don't know how history is taught," etc.*: Shannon, "Development of Communication and Computing, and My Hobby."

Chapter 31: *The Illness*

268 *"She is leaving him"*: Debra Dean, *The Madonnas of Leningrad* (New York: Harper Perennial, 2007), 119.

268 *"Claude was never a person"*: Robert Gallager, interviewed by the authors, August 8, 2014.

269 *"Do you juggle?"*: Peggy Shannon, interviewed by the authors, December 9, 2015.

269 *"In 1983, he went to a doctor"*: Betty Shannon, interviewed by the authors, November 12, 2015.

269 *"very quiet"*: Quoted in "Claude Shannon: Father of the Information Age."

269 *"They felt like they'd earned"* . . . *"She was the primary caretaker"* . . . *"There were days"*: Peggy Shannon, interviewed by the authors, December 9, 2015.

270 *"I asked him something"*: Robert Fano, interviewed by the authors, October 23, 2015.

270 *"Oddly enough"*: Betty Shannon, interviewed by the authors, November 12, 2015.

270 *speed finally approached, but did not break, the Shannon Limit*: These were the "turbo codes" discovered by Claude Berrou and colleagues. Codes capable of a similar speed had been discovered by Robert Gallager in 1960, but "the decoding process he proposed was simply too complicated for 1960s-era technology." See Øyvind Ytrehus, "An Introduction to Turbo Codes and Iterative Decoding," *Telektronikk* 98, no. 1 (2002): 65–78; Larry Hardesty, "Explained: Gallager Codes," *MIT News*, January 21, 2010, news.mit.edu/2010/gallager-codes-0121.

270 *"The sides of his personality"*: Peggy Shannon, interviewed by the authors, December 9, 2015.

270 *"The last time I saw Claude"*: Arthur Lewbel, interviewed by the authors, August 8, 2014.

270 *"home was a real refuge" . . . "she was very devoted"*: Peggy Shannon, interviewed by the authors, December 9, 2015.

271 *"at noon I'd go over" . . . "He still liked"*: Betty Shannon, interviewed by the authors, November 12, 2015.

Chapter 32: *Aftershocks*

273 *"The true spirit of delight"*: Bertrand Russell, "The Study of Mathematics," in *Mysticism and Logic and Other Essays* (London: Longman, 1919).

273 *"A uniquely playful" . . . "Shannon radiated"*: "Remembering Claude Shannon."

274 *"the logical and natural development"*: Ibid.

274 *"[His] ideas form a beautiful symphony"*: Ibid.

274 *"It was like an earthquake"*: Anthony Ephremides, interviewed by the authors, May 31, 2016.

275 *"For many scientists"*: Michael Urheber. *Bava's Gift: Awakening to the Impossible.*

275 *"all the advanced signal processing"*: Quoted in "Claude Shannon: Father of the Information Age."

276 *"In our age"*: Quoted in Mark Semenovich Pinsker, "Reflections of Some Shannon Lecturers," *IEEE Information Theory Society Newsletter*, Summer 1998, 22.

277 *"For him, the harder a problem"*: George Johnson, "Claude Shannon, Mathematician, Dies at 84," *New York Times*, February 27, 2001.

277 *"Courage is one of the things"*: Richard Hamming, "You and Your Research," lecture, Bell Communications Research Colloquium Seminar, March 7, 1986, www.cs.virginia.edu/~robins/YouAndYourResearch.html.

278 *"When you work with someone like Shannon"*: Leonard Kleinrock, interviewed by the authors, September 16, 2016.

279 *"I've been more interested"*: Shannon, interviewed by John Horgan (unpublished).

279 *"He was not interested"*: Henry Pollak, interviewed by the authors, August 7, 2014.

279 *"Shannon's puzzle-solving"*: Robert Gallager, "The Impact of Information Theory on Information Technology," lecture slides, February 28, 2006.

280 *"Dear Dennis"*: Quoted in John Horgan, "Poetic Masterpiece of Claude Shannon, Father of Information Theory, Published for the First Time," *Scientific American*, March 28, 2011, blogs.scientificamerican.com/cross-check/poetic-masterpiece-of-claude-shannon-father-of-information-theory-published-for-the-first-time.

Acknowledgments

283 *"Modern man"*: Arthur Koestler, *The Act of Creation* (London: Hutchinson, 1976), 264.

Bibliography

Books and Articles

Aftab, Omar, et al. "Information Theory and the Digital Age." web.mit.edu/6.933/www/Fall2001/Shannon2.pdf.

Allen, Garland E. "The Eugenics Record Office at Cold Spring Harbor: An Essay in Institutional History." *Osiris* 2, no. 2 (1986): 225–64.

Anderson, Benedict. *Imagined Communities: Reflections on the Origin and Spread of Nationalism*. Rev. ed. New York: Verso, 2006.

Aspray, William. "The Scientific Conceptualization of Information: A Survey." *IEEE Annals of the History of Computing* 7, no. 2 (1985): 117–40.

"The Atlantic Telegraph Expedition." *Times* (London), July 15, 1858.

Auden, W. H. "Foreword." In Dag Hammarskjöld, *Markings*. New York: Knopf, 1964.

Bandara, Lashi. "Explainer: The Point of Pure Mathematics." *The Conversation,* August 1, 2011. theconversation.com/explainer-the-point-of-pure-mathematics-2385.

Barzman, Norma. *The Red and the Blacklist: The Intimate Memoir of a Hollywood Expatriate*. New York: Nation Books, 2003.

Beauzamy, Bernard. "Real Life Mathematics." Lecture, Dublin Mathematical Society, February 2001. scmsa.eu/archives/BB_real_life_maths_2001.htm.

Beek, Peter J., and Arthur Lewbel. "The Science of Juggling." *Scientific American* 273, no. 5 (November 1995): 92–97.

Bello, Francis. "The Information Theory." *Fortune,* December 1953, 136–158.

———. "The Young Scientists." *Fortune,* June 1954, 142–48.

Blackman, R. B., H. W. Bode, and C. E. Shannon. "Data Smoothing and Prediction in Fire-Control Systems." Summary Technical Report of Division 7, NDRC, Volume I: Gunfire Control, ed. Harold Hazen. Washington, DC: Office of Scientific Research and Development, National Defense Research Committee, 1946.

Branford, Benchara. *A Study of Mathematical Education*. Oxford: Clarendon, 1908.

Brewer, Brock. "The Man-Machines May Talk First to Dr. Shannon." *Vogue,* April 15, 1963, 139.

"A Brief History of Gaylord Community Schools—1920 to 1944." *Otsego County Herald Times*, May 2, 1957. goo.gl/oVb0pT.

Brown, Anne S. "Historical Study: The National Security Agency Scientific Advisory Board, 1952–1963." Washington, DC: NSA Historian, Office of Central Reference, 1965.

Brueggeman, Brenda Jo. *Deaf Subjects: Between Identities and Places*. New York: New York University Press, 2009.

Burke, Colin B. *It Wasn't All Magic: The Early Struggle to Automate Cryptanalysis, 1930s–1960s*. United States Cryptologic History, Special Series, Volume 6. Center for Cryptologic History. Washington, DC: National Security Agency, 2002.

Burks, Frances Williston. *Barbara's Philippine Journey*. Yonkers-on-Hudson, NY: World Book, 1921.

Bush, Vannevar. "As We May Think." *Atlantic*, July 1945.

———. *Pieces of the Action*. New York: Morrow, 1970.

Carter, Samuel. *Cyrus Field: Man of Two Worlds*. New York: Putnam, 1968.

Cerny, Melinda. "Engineering Industry Honors Shannon, His Hometown." *Otsego Herald Times*, September 3, 1998.

Chiu, Eugene, et al. "Mathematical Theory of Claude Shannon." December 2001. web.mit.edu/6.933/www/Fall2001/Shannon1.pdf.

Clarke, Arthur C. *Voice Across the Sea: The Story of Deep Sea Cable-Laying, 1858–1958*. London: Muller, 1958.

"Claude Shannon Demonstrates Machine Learning." Bell Laboratories, 2014. www.youtube.com/watch?v=vPKkXibQXGA.

"Claude Shannon: Father of the Information Age." University of California Television, 2002. www.youtube.com/watch?v=z2Whj_nL-x8.

Clymer, A. Ben. "The Mechanical Analog Computers of Hannibal Ford and William Newell." *IEEE Annals of the History of Computing* 15, no. 2 (1993): 19–34.

Cocks, James Fraser, and Cathy Abernathy. *Pictorial History of Ann Arbor, 1824–1974*. Ann Arbor: Michigan Historical Collections/Bentley Historical Library Ann Arbor Sesquicentennial Committee, 1974.

Conway, Flo, and Jim Siegelman. *Dark Hero of the Information Age: In Search of Norbert Wiener, The Father of Cybernetics*. New York: Basic, 2005.

Cook, Gareth. "The Singular Mind of Terry Tao." *New York Times*, July 24, 2015.

Coughlin, Kevin. "Claude Shannon: The Genius of the Digital Age." *Star-Ledger* (New Jersey), February 28, 2001.

Crawford, Matthew. *Shop Class as Soulcraft: An Inquiry into the Value of Work*. New York: Penguin, 2010.

Crow, James F. "Shannon's Brief Foray Into Genetics." *Genetics* 159, no. 3 (2001): 915–17.

Davenport, C. B. *Naval Officers: Their Heredity and Development*. Washington, DC: Carnegie Institution of Washington, 1919.

Dean, Debra. *The Madonnas of Leningrad*. New York: Harper Perennial, 2007.

De Cogan, Donard. "Dr. E.O.W. Whitehouse and the 1858 Trans-Atlantic Cable." *History of Technology* 10 (1985): 1–15.

De Rosa, L. A. "In Which Fields Do We Graze?" *IRE Transactions on Information Theory* 1, no. 3 (1955): 2.

Dembart, Lee. "Book Review: Putting on Thinking Caps Over Artificial Intelligence." *Los Angeles Times*, August 15, 1989.

Diaconis, Persi, and Ron Graham. *Magical Mathematics*. Princeton, NJ: Princeton University Press, 2012.

Doob, J. L. "Review of *A Mathematical Theory of Communication*." *Mathematical Review* 10 (1949): 133.

Dunkel, Otto, H. L. Olson, and W. F. Cheney, Jr. "Problems and Solutions." *American Mathematical Monthly* 41, no. 3 (March 1934): 188–89.

"Enrico Rastelli." *Vanity Fair*, February 1932, 49.

Ephremides, Anthony. "Claude E. Shannon 1916–2001." *IEEE Information Theory Society Newsletter*, March 2001.

Feynman, Richard P. *Surely You're Joking, Mr. Feynman*. Reprint ed. New York: Norton, 1997.

Fisher, Lawrence. "Bernard M. Oliver Is Dead at 79; Led Hewlett-Packard Research." *New York Times,* November 28, 1995.

Freeman, John Ripley. "Study No. 7 for New Buildings for the Massachusetts Institute of Technology." MIT Libraries, Institute Archives and Special Collections. libraries.mit.edu/archives/exhibits/freeman.

Freudenthal, Hans. "Norbert Wiener." In *Complete Dictionary of Scientific Biography*. www.encyclopedia.com/people/science-and-technology/mathematics-biographies/norbert-wiener

Friedman, Norman. *Naval Firepower: Battleship Guns and Gunnery in the Dreadnought Era*. Barnsley, England: Seaforth, 2008.

Frize, Monique, Peter Frize, and Nadine Faulkner. *The Bold and the Brave*. Ottawa, Canada: University of Ottawa Press, 2009.

Fry, Thornton C. "Industrial Mathematics." *Bell Systems Technical Journal* 20, no. 3 (July 1941): 255–92.

Fussell, Paul. *Class: A Guide Through the American Status System*. Reissue ed. New York: Touchstone, 1992.

Gallager, Robert G. "Claude E. Shannon: A Retrospective on His Life, Work, and Impact." *IEEE Transactions on Information Theory* 47, no. 7 (2001): 2681–95.

———. "The Impact of Information Theory on Information Technology." Lecture slides. February 28, 2006.

"Gaylord Locals." *Otsego County Herald Times*, November 15, 1934.

"Gaylord's Claude Shannon: 'Einstein of Mathematical Theory.'" *Gaylord Herald Times*, October 11, 2000.

Gertner, Jon. *The Idea Factory: Bell Labs and the Great Age of American Innovation*. New York: Penguin, 2012.

Gifford, Walter. "The Prime Incentive." *Bell Laboratories Records.* Vols. 1 and 2. September 1925–September 1926.

Gleick, James. *The Information: A History, a Theory, a Flood*. New York: Pantheon, 2011.

Golomb, Solomon W. "Claude Elwood Shannon." *Notices of the AMS* 49, no. 1 (2001): 8–10.

———. "Retrospective: Claude E. Shannon (1916–2001)." *Science,* April 20, 2001.

Graham, C. Wallace, et al., eds. *1934 Michiganensian*. Ann Arbor, Michigan, 1934.

Guizzo, Erico Marui. "The Essential Message: Claude Shannon and the Making of Information Theory." MS diss., Massachusetts Institute of Technology, 2003.

Hamming, Richard. "You and Your Research." Lecture, Bell Communications Research Colloquium Seminar, March 7, 1986. www.cs.virginia.edu/~robins/YouAndYourResearch.html.

Hapgood, Fred. *Up the Infinite Corridor: MIT and the Technical Imagination*. New York: Basic Books, 1994.

Hardesty, Larry. "Explained: Gallager Codes." *MIT News,* January 21, 2010. news.mit.edu/2010/gallager-codes-0121.

Hardy, G. H. *A Mathematician's Apology.* Cambridge: Cambridge University Press, 2013.

Harpster, Jack. *John Ogden, The Pilgrim (1609–1682): A Man of More than Ordinary Mark*. Cranbury, NJ: Associated University Presses, 2006.

Hartley, Ralph. "Transmission of Information," *Bell System Technical Journal* 7, no. 3 (July 1928): 535–63.

Hartree, D. R. "The Bush Differential Analyzer and Its Applications." *Nature* 146 (September 7, 1940): 319–23.

Hatch, David A., and Robert Louis Benson. "The Korean War: The SIGINT Background." National Security Agency. www.nsa.gov/public_info/declass/korean_war/sigint_bg.shtml.

Hodges, Andrew. *Alan Turing: The Enigma*. Princeton, NJ: Princeton University Press, 1983.

———. "Alan Turing as UK-USA Link, 1942 Onwards." Alan Turing Internet Scrapbook. www.turing.org.uk/scrapbook/ukusa.html.

Horgan, John. "Claude E. Shannon: Unicyclist, Juggler, and Father of Information Theory." *Scientific American,* January 1990.

———. "Poetic Masterpiece of Claude Shannon, Father of Information Theory, Published for the First Time." *Scientific American,* March 28, 2011. blogs.scientificamerican.com/cross-check/poetic-masterpiece-of-claude-shannon-father-of-information-theory-published-for-the-first-time/.

Hunt, Bruce J. "Scientists, Engineers, and Wildman Whitehouse: Measurement and Credibility in Early Cable Telegraphy." *British Journal for the History of Science* 29, no. 2 (1996): 155–69.

Inamori, Kazuo. "Philosophy." Inamori Foundation, April 12, 1984. www.inamori-f.or.jp/en/kyoto_prize/.

"Institute Reports on Claude Shannon." *Otsego County Herald Times,* February 8, 1940.

INTOSAI Standing Committee on IT Audit. "1 + 1 = 1: A Tale of Genius." *IntoIT* 18 (2003): 52–57.

Isaacson, Walter. *The Innovators: How a Group of Inventors, Hackers, Geniuses, and Geeks Created the Digital Revolution*. New York: Simon & Schuster, 2014.

Jain, Naresh. "Record of the Celebration of the Life of Joseph Leo Doob." www.math.uiuc.edu/People/doob_record.html.

Jerison, David, I. M. Singer, and Daniel W. Stroock, eds. *The Legacy of Norbert Wiener: A Centennial Symposium in Honor of the 100th Anniversary of Norbert Wiener's Birth, October 8–14, 1994, Massachusetts Institute of Technology, Cambridge, Massachusetts*. Providence, RI: American Mathematical Society, 1997.

Johnson, George. "Claude Shannon, Mathematician, Dies at 84." *New York Times*, February 27, 2001.

———. *Fire in the Mind*. New York: Vintage, 1995.

Johnson, W. E. "The Logical Calculus." *Mind: A Quarterly Review of Psychology and Philosophy* 1 (1892): 3–30, 235–50, 340–57.

Kahn, David. *The Codebreakers: The Story of Secret Writing*. New York: Macmillan, 1953.

———. *How I Discovered World War II's Greatest Spy and Other Stories of Intelligence and Code*. Boca Raton, FL: Auerbach, 2014.Kahn, Robert E. "A Tribute to Claude E. Shannon." *IEEE Communications Magazine*, July 2001, 18–22.

Kaplan, Fred. "Scientists at War." *American Heritage* 34, no. 4 (June 1983): 49–64.

Kettlewell, Julie. "Gaylord Honors 'Father to the Information Theory.'" *Otsego Herald Times*, September 3, 1998.

Kimball, Warren F., ed. *Churchill and Roosevelt: The Complete Correspondence*. Vol. 3. Princeton, NJ: Princeton University Press, 1984.

Kipling, Rudyard. "The Deep Sea Cables." In *Rudyard Kipling's Verse*. Garden City, NY: Doubleday, Page, 1922.

Kline, Ronald R. *The Cybernetics Moment: Or Why We Call Our Age the Information Age*. Baltimore: Johns Hopkins University Press, 2015.

Koestler, Arthur. *The Act of Creation*. London: Hutchinson, 1976.

Lewbel, Arthur. "A Personal Tribute to Claude Shannon." www2.bc.edu/~lewbel/Shannon.html.

Lewes, George Henry. *The Principles of Success in Literature*. Berkeley: University of California Press, 1901.

Livingston, G. R. "Problems for Solution." *American Mathematical Monthly* 41, no. 6 (June 1934): 390.

Lucky, Robert W. *Silicon Dreams: Information, Man, and Machine*. New York: St. Martin's, 1991.

MacKay, David J. C. *Information Theory, Inference, and Learning Algorithms* Cambridge: Cambridge University Press, 2003.

Massey, James L. "Information Theory: The Copernican System of Communications." *IEEE Communications Magazine* 22, no. 12 (1984): 26–28.

McEliece, R. J. *The Theory of Information and Coding: Student Edition*. New York: Cambridge University Press, 2004.

Merzbach, Uta C., and Carl B. Boyer. *A History of Mathematics*. 3rd ed. Hoboken, NJ: John Wiley & Sons, 2011.

Minck, John. "Inside HP: A Narrative History of Hewlett-Packard from 1939–1990." www.hpmemoryproject.org/timeline/john_minck/inside_hp_03.htm.

Mindell, David A. "Automation's Finest Hour: Bell Labs and Automatic Control in WWII." *IEEE Control Systems* 15 (1995): 72–80.

———. *Between Human and Machine: Feedback, Control, and Computing before Cybernetics*. Baltimore: Johns Hopkins University Press, 2002.

"MIT Professor Claude Shannon dies; Was Founder of Digital Communications." *MIT News,* February 27, 2001. newsoffice.mit.edu/2001/shannon.

Mitchell, Silas Weir. "The Last of a Veteran Chess Player." *Chess Monthly,* 1857.

Morse, Philip McCord. *In at the Beginnings: A Physicist's Life.* Cambridge, MA: MIT Press, 1977.

Moulton-Barrett, Maria. *Graphotherapy.* New York: Trafford, 2005.

"Mouse with a Memory." *Time,* May 19, 1952, 59–60.

"Mrs. Mabel Shannon Dies in Chicago," *Otsego County Herald Times,* December 27, 1945.

Nahin, Paul J. *The Logician and the Engineer: How George Boole and Claude Shannon Created the Information Age.* Princeton, NJ: Princeton University Press, 2013.

Nasar, Sylvia. *A Beautiful Mind: The Life of Mathematical Genius and Nobel Laureate John Nash.* New York: Simon & Schuster, 1998.

National Register of Historic Places application. Edmund Dwight House. Massachusetts Cultural Resource Information System. mhc-macris.net/Details.aspx?MhcId=WNT.19.

Norberg, Arthur L. "An Interview with Bernard More Oliver." Charles Babbage Institute for the History of Information Processing, August 9, 1985.

"NSA Regulation Number 11-3." National Security Agency, January 22, 1953. ia601409.us.archive.org/16/items/41788579082758/41788579082758.pdf.

Nyquist, Harry. "Certain Factors Affecting Telegraph Speed." *Bell System Technical Journal* (April 1924): 324–46.

———. "Certain Topics in Telegraph Transmission Theory." *Transactions of the AIEE* 47 (April 1928): 617–44.

"Obituary: Thornton Carl Fry." American Astronomical Society, January 1, 1991.

Oliver, B., J. Pierce, and C. Shannon. "The Philosophy of PCM." *Proceedings of the IRE* 36, no. 11 (November 1948): 1324–31.

O'Neill, Bradley. "Dead Medium: The Comparator; the Rapid Selector." www.deadmedia.org/notes/1/017.html.

Owens, F. W., and Helen B. Owens. "Mathematics Clubs—Junior Mathematics Club, University of Michigan." *American Mathematical Monthly* 43, no. 10 (December 1936): 636.

Owens, Larry. "Vannevar Bush and the Differential Analyzer: The Text and Context of an Early Computer." *Technology and Culture* 27, no. 1 (1986): 63–95.

Perkins, Thomas. *Valley Boy: The Education of Tom Perkins*. New York: Gotham, 2007.

Perry, John. *The Calculus for Engineers.* London: Edward Arnold, 1897.

Pierce, John. "Creative Thinking." Lecture. 1951.

———. "The Early Days of Information Theory." *IEEE Transactions on Information Theory* 19, no. 1 (1973): 3–8.

———. *An Introduction to Information Theory: Symbols, Signals, and Noise.* 2nd ed. New York: Dover, 1980.

Pinsker, Mark Semenovich. "Reflections of Some Shannon Lecturers." *IEEE Information Theory Society Newsletter* (Summer 1998): 22–23.

Platt, John. "Books That Make a Year's Reading and a Lifetime's Enrichment." *New York Times,* February 2, 1964.

Poe, Edgar Allan. "The Gold-Bug." In *The Gold-Bug and Other Tales*. Ed. Stanley Appelbaum. Mineola, NY: Dover, 1991.

———. "Maelzel's Chess Player." In *The Complete Tales of Edgar Allan Poe*. New York: Vintage Books, 1975.

Polster, Burkard. "The Mathematics of Juggling." qedcat.com/articles/juggling_survey.pdf.

Poundstone, William. *Fortune's Formula: The Untold Story of the Scientific Betting System That Beat the Casinos and Wall Street*. New York: Hill & Wang, 2005.

———. *How to Predict the Unpredictable: The Art of Outsmarting Almost Anyone*. New York: Oneworld, 2014.

Powers, Perry Francis, and H. G. Cutler. *A History of Northern Michigan and Its People*. Chicago: Lewis, 1912.

Price, Robert. "A Conversation with Claude Shannon: One Man's Approach to Problem Solving." *IEEE Communications Magazine* 22, no. 6 (May 1984): 123–26.

———. "Oral History: Claude E. Shannon." IEEE Global History Network, July 28, 1982. www.ieeeghn.org/wiki/index.php/Oral-History:Claude_E._Shannon.

Ratcliff, J. D. "Brains." *Collier's,* January 17, 1942.

Rees, Mina. "Warren Weaver." In National Academy of Sciences, *Biographical Memoirs,* vol. 57. Washington, DC: National Academy Press, 1987.

"Remembering Claude Shannon." Roy Rosenzweig Center for History and New Media, George Mason University, March–August 2001. chnm.gmu.edu/digitalhistory/links/cached/chapter6/6_19b_surveyresponse.htm.

Report of the Joint Committee to Inquire into the Construction of Submarine Telegraph Cables. London: Eyre & Spottiswoode, 1861.

Rheingold, Howard. *Tools for Thought*. Cambridge, MA: MIT Press, 2000.

Rosser, J. Barkley. "Mathematics and Mathematicians in World War II." In *A Century of Mathematics in America, Part 1*. Ed. Peter Duren. Providence, RI: American Mathematical Society, 1988.

Russell, Bertrand. "The Study of Mathematics." In *Mysticism and Logic and Other Essays*. London: Longman, 1919.

Sagan, Carl. *Pale Blue Dot: A Vision of the Human Future in Space*. New York: Random House, 1994.

Saxon, Wolfgang. "Albert G. Hill, 86, Who Helped Develop Radar in World War II." *New York Times*, October 29, 1996.

Schement, Jorge Reina, and Brent D. Ruben. *Between Communication and Information* 4. New Brunswick, NJ: Transaction, 1993.

Shannon, Claude Elwood. "The Bandwagon." *IRE Transactions—* Information *Theory* 2, no. 1 (1956): 3.

———. *Claude Elwood Shannon: Collected Papers*. Ed. N. J. A. Sloane and Aaron D. Wyner. New York: IEEE Press, 1992.

———. *Claude Shannon's Miscellaneous Writings*. Ed. N. J. A. Sloane and Aaron D. Wyner. Murray Hill, NJ: Mathematical Sciences Research Center, AT&T Bell Laboratories, 1993.

———. "Development of Communication and Computing, and My Hobby." Lecture, Inamori Foundation, Kyoto, Japan, November 1985. www.kyotoprize.org/wp/wp-content/uploads/2016/02/1kB_lct_EN.pdf.

———. "A Mathematical Theory of Communication." *Bell System Technical Journal* 27 (July, October 1948): 379–423, 623–56.

———. "Problems and Solutions—E58." *American Mathematical Monthly* 41, no. 3 (March 1934): 191–92.———. "A Symbolic Analysis of Relay and

Switching Circuits." *Transactions of the American Institute of Electrical Engineers* 57 (1938): 471–95.

———. "A Theorem on Color Coding." Bell Laboratories. Memorandum 40-130-153. July 8, 1940.

———. "The Use of the Lakatos-Hickman Relay in a Subscriber Sender." Bell Laboratories. Memorandum 40-130-179. August 3, 1940.

Sicilia, David B. "How the West Was Wired." *Inc.*, June 1997.

Snell, J. Laurie. "A Conversation with Joe Doob." 1997. www.dartmouth.edu/~chance/Doob/conversation.html.

Standage, Tom. *The Turk: The Life and Times of the Famous Eighteenth-Century Chess-Playing Machine*. New York: Walker, 2002.

"Step Back in Time: A New County Seat and the First Newspaper." *Gaylord Herald Times*, reprinted January 6, 2016.

Sterling, Christopher H. "Churchill and Intelligence—Sigsaly: Beginning the Digital Revolution." *Finest Hour 149* (Winter 2010–11): 31.

Sutin, Hillard A. "A Tribute to Mortimer E. Cooley." *Michigan Technic*, March 1935.

Sutton, R. M. "Problems for Solution." *American Mathematical Monthly* 40, no. 8 (October 1933): 491.

Thomson, Silvanus P. *The Life of William Thomson, Baron Kelvin of Largs*. London: Macmillan, 1910.

Thomson, William. "The Tides: Evening Lecture to the British Association at the Southampton Meeting, August 25, 1882." In Thomson, *Scientific Papers*, vol. 30. Ed. Charles W. Eliot. New York: Collier & Son, 1910.

Thorp, Edward O. "The Invention of the First Wearable Computer." *Proceedings of the 2nd IEEE International Symposium on Wearable Computers*, October 1998, 4–8.

Tompkins, Dave. *How to Wreck a Nice Beach: The Vocoder from World War II to Hip-Hop: The Machine Speaks*. Chicago: Stop Smiling Books, 2011.

Trew, Delbert. "Barbed Wire Telegraph Lines Brought Gossip and News to Farm and Ranch." *Farm Collector,* September 2003.

Tribus, Myron, and Edward C. McIrving. "Energy and Information." *Scientific American* 225 (1971): 179–88.

Turing, Alan. "Alan Turing's Report from Washington DC, November 1942."

Van den Herik, H. J. "An Interview with Claude Shannon (September 25, 1980 in Linz, Austria)." *ICCA Journal* 12, no. 4 (1989): 221–26.

"Vannevar Bush: General of Physics." *Time,* April 3, 1944.

Von Foerster, Heinz, Margaret Mead, and Hans Lukas Teuber, eds. *Cybernetics: Transactions of the Eighth Conference March 15–16, 1951*. New York: Josiah Macy, Jr. Foundation, 1952.

Von Neumann, John. "First Draft of a Report on the EDVAC." In *The Origins of Digital Computers: Selected Papers*. Ed. Brian Randell. New York: Springer-Verlag, 1973.

Waldrop, W. Mitchell. "Claude Shannon: Reluctant Father of the Digital Age." *MIT Technology Review,* July 1, 2001. www.technologyreview.com/s/401112/claude-shannon-reluctant-father-of-the-digital-age.

Weaver, Warren. "Careers in Science." In *Listen to Leaders in Science*. Ed. Albert Love and James Saxon Childers. Atlanta: Tupper & Love/David McKay, 1965.

———. "Four Pieces of Advice to Young People." In *The Project Physics Course Reader: Concepts of Motion*. Ed. Gerald Holton et al. New York: Holt, Rinehart & Winston, 1970.

———. *Science and Imagination: Selected Papers*. New York: Basic Books, 1967.

Weyl, Hermann. *Space—Time—Matter*. 4th ed. Trans. Henry L. Brose. New York: Dover, 1950.

Whaland, Norman. "A Computer Chess Tutorial." *Byte,* October 1978, 168–81.

Whitehouse, E. O. Wildman. "The Law of Squares—Is It Applicable or Not to the Transmission of Signals in Submarine Circuits?" *Athenaeum*, August 30, 1856, 1092–93.

———. "Report on a Series of Experimental Observations on Two Lengths of Submarine Electric Cable, Containing, in the Aggregate, 1,125 Miles of Wire, Being the Substance of a Paper Read Before the British Association for the Advancement of Science, at Glasgow, Sept. 14th, 1855." Brighton, England, 1855.

"Who We Are." Douglass Residential College, Rutgers University. douglass.rutgers.edu/history.

Wiener, Norbert. *Cybernetics, or Control and Communication in the Animal and the Machine*. 2nd ed. Cambridge, MA: MIT Press, 1961.

———. *Ex-Prodigy: My Childhood and Youth*. Cambridge, MA: MIT Press, 1964.

Wildes, Karl L., and Nilo A. Lindgren. *A Century of Electrical Engineering and Computer Science at MIT, 1882–1982*. Cambridge, MA: MIT Press, 1986.

Wilson, Philip K. "Harry Laughlin's Eugenic Crusade to Control the 'Socially Inadequate' in Progressive Era America." *Patterns of Prejudice* 36, no. 1 (2002): 49–67.

Wittgenstein, Ludwig. *Philosophical Investigations*. Trans. G. E. M. Anscombe et al. Ed. P. M. S. Hacker and Joachim Schulte. 4th ed. Malden, MA: Blackwell, 2009.

"Youthful Instructor Wins Noble Award." *New York Times,* January 24, 1940.

Ytrehus, Øyvind. "An Introduction to Turbo Codes and Iterative Decoding." *Telektronikk* 98, no. 1 (2002): 65–78.

Zachary, G. Pascal. *Endless Frontier: Vannevar Bush, Engineer of the American Century*. Cambridge, MA: MIT Press, 1999.

Zorpette, Glenn. "Parkinson's Gun Director." *IEEE Spectrum* 26, no. 4 (1989): 43.

Archival Material

Claude Elwood Shannon Papers. Library of Congress. Washington, DC.

Claude Shannon Alumnus File. Bentley Historical Library. University of Michigan. Ann Arbor, MI.

Claude Shannon Alumnus File. Seeley Mudd Library. Princeton University. Princeton, NJ.

Institute of Communications Research Records. University of Illinois Archives. Urbana, IL.

Kelvin Collection. University of Glasgow. Glasgow, Scotland.

Office of the President Records. MIT Archive. Cambridge, MA.

Vannevar Bush Papers. Library of Congress. Washington, DC.

Warren S. McCulloch Papers. American Philosophical Society. Philadelphia, PA.

Interviews

Barzman, Norma. Interviewed by the authors, December 21, 2014.

Ephremides, Anthony. Interviewed by the authors, May 31, 2016.

Fano, Robert. Interviewed by the authors, October 23, 2015.

Fry, Thornton C. Interviewed by Deirdre M. La Porte, Henry O. Pollak, and G. Baley Price, January 3–4, 1981.

Gallager, Robert. Interviewed by the authors, August 8, 2014.

Graham, Ronald. Interviewed by the authors, August, 23, 2014.

Greenberger, Martin. Interviewed by the authors, May 5, 2016.

Jacobs, Irwin. Interviewed by the authors, January 1, 2015.

Kailath, Thomas. Interviewed by the authors, June 2, 2016.

Kleinrock, Leonard. Interviewed by the authors, September 16, 2016.

Lewbel, Arthur. Interviewed by the authors, August 8, 2014.

McMillan, Brockway. Interviewed by the authors, January 4, 2016.

Moulton-Barrett, Maria. Interviewed by the authors, January 2, 2015, and January 21, 2016.

Pollak, Henry. Interviewed by the authors, August 7, 2014.

Roberts, Larry. Interviewed by the authors, September 26, 2016.

Shannon, Betty. Interviewed by the authors, November 12, 2015.

Shannon, Claude. Interviewed by Friedrich-Wilhelm Hagemeyer, February 28, 1977.

Shannon, Claude, and Betty Shannon. Interviewed by Donald J. Albers, 1990.

Shannon, Peggy. Interviewed by the authors, December 9, 2015.

Thorp, Edward. Interviewed by the authors, March 21, 2016.

Verdú, Sergio. Interviewed by the authors, September 6, 2015.

Illustration Credits

Otsego County Times: 1

Necrology files, Bentley Historical Library, University of Michigan: 2

US Patent and Trademark Office, Patent No 407,130: 3

College of Engineering Records, Bentley Historical Library, University of Michigan: 4

Detroit Publishing Company Photograph Collection: 5

Shannon family: 6, 9, 18, 23, 24, 25, 27, 29, 33, 34, 36, 42, 43, 44, 45

Roger W / Flickr: 7

The World's Work (1902): 8

Computer Laboratory, University of Cambridge. Reproduced by permission: 10

Courtesy MIT Museum: 11, 22, 31, 32, 35, 37, 38, 40

Courtesy of Harry H. Laughlin Papers, Special Collections Department, Pickler Memorial Library, Truman State University: 12

Photographer unknown. From the Shelby White and Leon Levy Archives Center, Institute for Advanced Study, Princeton, NJ, USA: 13

Public Domain: 14, 19

Nokia Bell Labs: 15, 16, 21, 26

NASA: 17

Courtesy of the Hartley Family / Wikimedia Commons: 20

Library of Congress: 28

Alfred Eisenstaedt / The LIFE Picture Collection / Getty Images: 30

Courtesy of Edward Thorp: 39

Steve Hansen / The LIFE Images Collection / Getty Images: 41

Index

About the Authors

Jimmy Soni has served as an editor at the *New York Observer* and the *Washington Examiner*, and as managing editor of the *Huffington Post*. He is a former speechwriter, and his written work and commentary have appeared in *Slate, The Atlantic*, and CNN, among other outlets. He is a graduate of Duke University.

Rob Goodman is a doctoral candidate at Columbia University and a former congressional speechwriter. He has written for *Slate, The Atlantic, Politico,* and *The Chronicle of Higher Education*. His scholarly work has appeared in *History of Political Thought*, the *Kennedy Institute of Ethics Journal*, and *The Journal of Medicine and Philosophy.*

Soni and Goodman are the coauthors of *Rome's Last Citizen: The Life and Legacy of Cato, Mortal Enemy of Caesar.*